2nd Edition

BIOLOGY

W9-AWF-664

SUPER REVIEW®

By the Staff of
Research & Education Association

Research & Education Association
Visit our website at: www.rea.com

Research & Education Association
61 Ethel Road West
Piscataway, New Jersey 08854
E-mail: info@rea.com

BIOLOGY
SUPER REVIEW®

Printed in the United States of America

Library of Congress Control Number 2013932759

ISBN-13: 978-0-7386-1121-1
ISBN-10: 0-7386-1121-2

REA's *Biology Super Review*®

Need help with Biology? Want a quick review or refresher for class? This is the book for you!

REA's *Biology Super Review*® gives you everything you need to know!

This *Super Review*® can be used as a supplement to your high school or college textbook, or as a handy guide for anyone who needs a fast review of the subject.

- **Comprehensive yet concise coverage** – review covers the material that is typically taught in a beginning-level biology course. Each topic is presented in a clear and easy-to-understand format that makes learning easier.

- **Questions and answers for each topic** – let you practice what you've learned and build your biology skills.

- **End-of-chapter quizzes** – gauge your understanding of the important information you need to know, so you'll be ready for any biology assignment, quiz, or test.

Whether you need a quick refresher on the subject, or are prepping for your next test, we think you'll agree that REA's *Super Review*® provides all you need to know!

Available Super Review® Titles

ARTS/HUMANITIES
Basic Music
Classical Mythology
History of Architecture
History of Greek Art

BUSINESS
Accounting
Macroeconomics
Microeconomics

COMPUTER SCIENCE
C++
Java

HISTORY
Canadian History
European History
United States History

LANGUAGES
English
French
French Verbs
Italian
Japanese for Beginners
Japanese Verbs
Latin
Spanish

MATHEMATICS
Algebra & Trigonometry
Basic Math & Pre-Algebra
Calculus
Geometry
Linear Algebra
Pre-Calculus
Statistics

SCIENCES
Anatomy & Physiology
Biology
Chemistry
Entomology
Geology
Microbiology
Organic Chemistry I & II
Physics

SOCIAL SCIENCES
Psychology I & II
Sociology

WRITING
College & University Writing

About Research & Education Association

Founded in 1959, Research & Education Association is dedicated to publishing the finest and most effective educational materials— including study guides and test preps—for students in middle school, high school, college, graduate school, and beyond.

Today, REA's wide-ranging catalog is a leading resource for teachers, students, and professionals. Visit *www.rea.com* to see a complete listing of all our titles.

Acknowledgments

We would like to thank Pam Weston, Publisher, for setting the quality standards for production integrity and managing the publication to completion; Larry B. Kling, Vice President, Editorial, for his supervision of revisions and overall direction; Kelli Wilkins, Copywriter, for coordinating development of this edition; and Christine Saul, Senior Graphic Designer, for designing our cover.

Contents

CHAPTER 1

The Chemical and Molecular Basis of Life

1.1 The Elements

Element – An element is a substance which cannot be decomposed into simpler or less complex substances by ordinary chemical means.

Compound – Compounds are a combination of elements present in definite proportions by mass. These are substances which can be decomposed by chemical means.

Mixtures – Mixtures contain two or more substances, each of which retains its original properties and can be separated from the others by relatively simple means. They do not have a definite composition.

Atoms – Each element is made up of one kind of atom. An atom is the smallest part of an element which can combine with other elements. Each atom consists of:

   A) **Atomic Nucleus** – small, dense center of an atom.

   B) **Proton** – positively charged particle of the nucleus.

   C) **Neutron** – electrically neutral particle of the nucleus.

   D) **Electron** – negatively charged particle which moves around the nucleus.

In normal, neutral atoms, the number of electrons is equal to the number of protons.

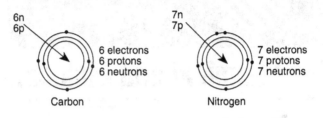

6n
6p
6 electrons
6 protons
6 neutrons

Carbon

7n
7p
7 electrons
7 protons
7 neutrons

Nitrogen

Figure 1.1 Atomic Structure of carbon and nitrogen

Atomic Weight – The total number of protons and neutrons in a nucleus is the atomic weight (mass number). This number approximates the total mass of the nucleus.

Atomic Number – The atomic number is equal to the number of protons in the nucleus of an element.

Isotope – Atoms of the same element that have a different number of neutrons are known as isotopes. All isotopes of the same element have essentially the same chemical properties but their physical properties may be affected.

Ions – Atoms or groups of atoms which have lost or gained electrons are called ions. One of the ions formed is always electrically positive and the other electrically negative.

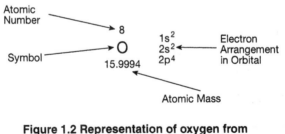

Atomic
Number

8

$1s^2$
$2s^2$
$2p^4$

Electron
Arrangement
in Orbital

Symbol

O

15.9994

Atomic Mass

**Figure 1.2 Representation of oxygen from
Periodic Table of the Elements**

Problem Solving Examples:

Q Define the following terms: atom, isotope, ion. Could a single particle of matter be all three simultaneously?

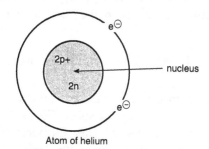

Atom of helium

A An atom is the smallest particle of an element that can retain the chemical properties of that element. It is composed of a nucleus, which contains positively charged protons and neutral neutrons. Negatively charged electrons revolve in orbits around the nucleus. For example, a helium atom contains 2 protons, 2 neutrons, and 2 electrons.

An ion is an electrically charged atom or group of atoms. The electrical charge results when a neutral atom or group of atoms loses or gains one or more electrons during a chemical reaction. An ion which is negatively charged is called an anion, and a positively charged ion is called a cation.

Isotopes are alternate forms of the same element. An element is defined in terms of its atomic number, which is the number of protons in its nucleus. Isotopes of an element have the same number of protons, but a different number of neutrons. Since atomic mass is determined by the number of protons plus neutrons, isotopes of the same element have varying atomic masses. For example, deuterium (^2H) is an isotope of hydrogen, and has one neutron and one proton in its nucleus. Hydrogen has only a proton and no neutrons in its nucleus.

A single particle can be an atom, an ion, and an isotope simultaneously. The simplest example is the hydrogen ion (H^+). It is an atom which has lost one electron and thus developed a positive charge. Since it is charged, it is therefore an ion. A cation is a positively charged ion (i.e., H^+), and an anion is a negatively charged ion (i.e., Cl^-). If one compares the atomic number of H^+ (1) with that of deuterium (1), it is seen that although they have different atomic masses, since their atomic numbers are the same, they must be isotopes of one another.

 Describe the differences between an element and a compound.

 All substances are composed of matter, which means that they have mass and occupy space. Elements and compounds constitute two general classes of matter. Elements are substances that consist of identical atoms (i.e., atoms with identical atomic numbers). This definition of an element includes all isotopes of that element. Hence O^{18} and O^{16} are both considered to be elemental oxygen. A compound is a substance that is composed of two or more different kinds of atoms (two or more different elements) combined in a definite weight ratio. This fixed composition of various elements, according to law of definite proportions, differentiates a compound from a mixture. Elements are the substituents of compounds. For example, water is a compound composed of the two elements hydrogen and oxygen in the ratio 2:1, respectively. This compound may be written as H_2O, which is the molecular formula of water. The subscript "2" that appears after the hydrogen (H) indicates that every molecule of water has two hydrogen atoms. There is no subscript after the oxygen (O) in the molecular formula of water, which indicates that there is only one oxygen atom per molecule of water. Hence, water is a compound whose molecules are each made up of two hydrogen atoms and one oxygen atom.

1.2 Chemical Bonds

Ionic Bond – This involves the complete transfer of an electron from one atom to another. Ionic bonds form between strong electron donors and strong electron acceptors. Ionic compounds are more stable than the individual elements.

Covalent Bond – This involves the sharing of pairs of electrons between atoms. Covalent bonds may be single, double, or triple.

Polar Covalent Bond – A polar covalent bond is a bond in which the charge is distributed asymmetrically within the bond.

Non-Polar Covalent Bond – A non-polar covalent bond is a bond where the electrons are distributed equally between two atoms.

Hydrogen Bond – A hydrogen bond is the attraction of a hydrogen atom, already covalently bonded to one electronegative atom, to a second electronegative atom of the same molecule or adjacent molecule. Usually these bonds are found in compounds that have strong electronegative atoms such as oxygen, fluorine, or nitrogen.

Van der Waals Forces – Van der Waals forces are weak linkages which occur between electrically neutral molecules or parts of molecules which are very close to each other.

Hydrophobic Interactions – Hydrophobic interactions occur between groups that are insoluble in water. These groups, which are non-polar, tend to clump together in the presence of water.

Problem Solving Examples:

 Distinguish between covalent and ionic bonds.

 A covalent bond is a bond in which two atoms are held together by a shared pair of electrons. An ionic bond is a bond in which oppositely charged ions are held together by electrical attraction.

In general, the electronegativity difference between two elements influences the character of their bond (see the table on the following page).

Electronegativity measures the relative ability of an atom to attract electrons in a covalent bond. Using Pauling's scale, where fluorine is arbitrarily given the value 4.0 units and other elements are assigned values relative to fluorine, an electronegativity difference of greater than 1.7 units is mostly ionic in character. Therefore, a bond between two atoms with an electronegativity difference of greater than 1.7 units is mostly ionic in character. If the difference is less than 1.7, the bond is predominantly covalent.

I A	II A	III A	IV A	V A	VI A	VII A
H						
2.1						
Li	Be	B	C	N	O	F
1.0	1.5	2.0	2.5	3.0	3.5	4.0
Na	Mg	Al	Si	P	S	Cl
0.9	1.2	1.5	1.8	2.1	2.5	3.0
K	Ca		Ge	As	Se	Br
0.8	1.0		1.8	2.0	2.4	2.8
Rb	Sr		Sn	Sb	Te	I
0.8	1.0		1.8	1.9	2.1	2.5
Cs	Ba		Pb	Bi		
0.7	0.9		1.7	1.8		

Electronegativities of main groups of elements

 What are hydrogen bonds? Describe fully the importance of hydrogen bonds in the biological world.

A hydrogen bond is a molecular force in which a hydrogen atom is shared between two atoms. Hydrogen bonds occur as a result of the uneven distribution of electrons in a polar bond. Here, the bonding electrons are more attracted to the highly electronegative oxygen atom, resulting in a slightly positive charge (δ^+) on the hydrogen and a slightly negative charge (δ^-) on the oxygen. A hydrogen bond is formed when the relatively positive hydrogen is attracted to a relatively negative atom of some other polarized bond. For example:

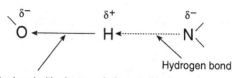

Hydrogen bond

Polar bond with electrons being attracted
to the more electronegative element, oxygen

The atom with which it forms the polar bond is called the hydrogen donor, while the other atom is the hydrogen acceptor. Note, however, that the bond is an electrostatic one – no electrons are shared or exchanged between the hydrogen and the negative dipole of the other molecule of the bond.

Hydrogen bonds are highly directional (note the arrows in the figure), and are strongest when all three atoms are colinear (when the bond angles between the atoms are 180°).

Bond energies of hydrogen bonds are in the range of about 3 to 7 kcal/mole. This is intermediate between the energy of a covalent bond and a van der Waals bond. However, only when the electronegative atoms are either F, O, or N is the energy of the bond enough to make it important.

Hydrogen bonds are responsible for the structure of water and its special properties as a biological solvent. There is extensive hydrogen bonding between water molecules, forming what is called the water matrix. The formation of the water matrix is what causes frozen H_2O (water) to float and be less dense than liquid form, sustaining life even in frozen ponds or rivers. This structure has profound effects on the freezing and boiling points of water and its solubility properties. Any molecule capable of forming a hydrogen bond can do so with water, which results in dissociation, or solubility of the molecule.

Hydrogen bonds are also most responsible for the maintenance of the shape of proteins. Since shape is crucial to protein function, this bonding is extremely important. For example, hydrogen bonds maintain the helical shape of keratin and collagen molecules and give them their characteristic strength and flexibility.

DNA helices are held together by hydrogen bonds. Bonding occurs between the base pairs. The intermediate bond strength of the hydrogen bond is ideal for the function of DNA – it is strong enough to give the molecule stability, yet weak enough to be broken with sufficient ease for replication and RNA synthesis.

 What are van der Waals forces? What is their significance in biological systems?

 Van der Waals forces are the weak attractive forces that molecules of non-polar compounds have for one another. These are the forces that allow non-polar compounds to liquefy and/or solidify. These forces are based on the existence of momentary dipoles within molecules of non-polar compounds. A dipole is the separation of opposite charges (positive and negative). A non-polar compound's average distribution of charge is symmetrical, so there is no net dipole. But, electrons are not static; they are constantly moving about. Thus, at any instant in time a small dipole will exist. This momentary dipole will affect the distribution of charge in nearby non-polar molecules, inducing charges. This induction happens because the negative end of the temporary dipole will repel electrons and the positive end attracts electrons. Thus, the neighboring non-polar molecules will have oppositely oriented dipoles:

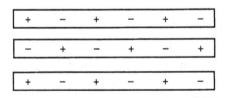

These momentary, induced dipoles are constantly changing, short range forces. But, their net result is attraction between molecules.

The attraction due to van der Waals forces steadily increases when two non-bonded atoms are brought closer together, reaching its maximum when they are just touching. Every atom has a van der Waals radius. If the two atoms are forced closer together than the minimum radius, van der Waals attraction is replaced by van der Waals repulsion (because of the positively charged nuclei). The atoms then try to return to equilibrium of the maximum radius.

Both attractive and repulsive van der Waals forces play important roles in many biological systems. These forces acting between non-polar chains of phospholipids serve to hold the membranes of living cells together.

1.3 Acids and Bases

Acid – An acid is a compound which dissociates in water and yields hydrogen ions [H+]. It is referred to as a proton donor.

Base – A base is a compound which dissociates in water and yields hydroxyl ions [OH-]. Bases are proton acceptors.

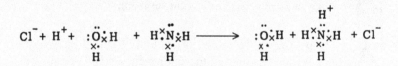

Figure 1.3. Reaction between hydrochloric acid (proton donor) and ammonia (proton acceptor)

pH – The degree of acidity or alkalinity is measured by pH.

$$pH = \frac{1}{\log[H^+]} = -\log[H^+]$$

pH = 7 → neutral
pH ‹ 7 → acidic
pH › 7 → basic

Problem Solving Examples:

 Differentiate between acids and bases. Give examples of each. How is water defined?

A There are essentially 2 widely used definitions of acids and bases: the Lowry-Bronsted definition and the Lewis definition. In the Lowry-Bronsted definition, an acid is a compound with the capacity to donate a proton, and a base is a compound with the capacity to accept a proton. In the Lewis definition, an acid has the ability to accept an electron pair and a base the ability to donate an electron pair.

Some common acids important to the biological system are acetic acid (CH_3COOH), carbonic acid (H_2CO_3), phosphoric acid (H_3PO_4), and water. Amino acids, the building blocks of protein, are compounds that contain an acidic group ($-COOH$). Some common bases are ammonia (NH_3), pyridine (C_5H_5N), and water. The nitrogenous bases important in the structure of DNA and RNA carry the purine or pyridine functional group. Water has the ability to act both as an acid ($H_2O \rightarrow H^+ + OH^-$) and as a base ($H_2O \rightarrow H^+ + H_3O^+$) depending on the conditions of the reaction, and is thus said to exhibit amphiprotic behavior.

 Q What does the "pH" of a solution mean?

A The pH (an abbreviation for "potential of hydrogen") of a solution is a measure of the hydrogen ion (H^+) concentration. Specifically, pH is defined as the negative log of the hydrogen ion concentration. A pH scale is used to quantify the relative acid or base strength. It is based upon the dissociation reaction of water: $H_2O \rightarrow H^+ + OH^-$. The dissociation constant (K) of this reaction is 1.0×10^{-14}. [H_2O] is the concentration of water (which is equal to one). The pH of water can be calculated from its dissociation constant K. Since one H^+ and one ^-OH are formed for every dissociated H_2O molecule, [H^+] = [^-OH].

$$1 \times 10^{-14} = [H^+][OH^-]; [H^+] = 1.0 \times 10^{-7}$$

$$pH = -\log [H^+] = -\log (1.0 \times 10^{-7}) = 7$$

A pH of 7 is considered to be neutral since there are equal concentrations of hydrogen and hydroxide ions (OH^-). The pH scale ranges from 0 to 14. Acidic compounds have a pH range of 0 to 7 and basic compounds have a range of 7 to 14.

1.4 Chemical Changes

Chemical Reaction - A chemical reaction refers to any process in which at least one chemical bond is either broken or formed. The outcome of a chemical reaction is a rearrangement of atoms and bonding patterns.

Laws of Thermodynamics

A) **First Law of Thermodynamics (Conservation of Energy)** – In any process, the sum of all energy changes must be zero.

B) **Second Law of Thermodynamics** – Any system tends toward a state of greater entropy, meaning randomness or disorder.

C) **The Third Law of Thermodynamics** – A perfect crystal, which is a completely ordered system, at absolute zero (0° Kelvin) would have perfect order, and therefore its entropy would be zero.

Stability of chemical system depends on:

A) **Enthalpy** – total energy content of a system.

B) **Entropy** – energy distribution.

Exergonic Reaction – Exergonic reactions release free energy; all spontaneous reactions are exergonic.

Endergonic Reaction – Endergonic reactions require the addition of free energy from an external source.

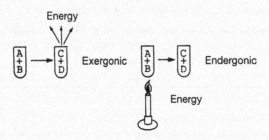

Figure 1.4 Exergonic and Endergonic Reactions

Problem Solving Example:

 What are the three laws of thermodynamics? Discuss the biological significance of the first two.

 The first law of thermodynamics states that energy can be converted from one form into another, but it cannot be created or destroyed. In other words, the energy of a closed system is constant. Thus, the first law is simply a statement of the law of conservation of energy; the sum of all energy changes must be zero.

The second law of thermodynamics states that the total entropy (a measure of the disorder or randomness of a system) of the universe is increasing. This is characterized by a decrease in the free energy, which is the energy available to do work. Thus, any spontaneous change that occurs (chemical, physical, or biological) will tend to increase the entropy of the universe.

The third law of thermodynamics refers to a completely ordered system, particularly, a perfect crystal. It states that a perfect crystal at absolute zero (0 Kelvin) would have perfect order, and therefore its entropy would be zero.

These three laws affect the biological as well as the chemical and physical worlds. Living cells do their work by using the energy stored in chemical bonds. The first law of thermodynamics states that every chemical bond in a given molecule contains an amount of energy equal to the amount that was necessary to link the atoms together. Thus, living cells turn other forms of energy into chemical bond energy and use it to do work. A living organism is a storehouse of potential chemical energy due to the many millions of atoms bonded together in each cell, so it might appear that the same energy could be passed continuously from organism to organism with no required extracellular energy source. However, the second law of thermodynamics tells us that every energy transformation results in a reduction in the usable or free energy of the system. Consequently, there is a steady increase in the amount of energy that is unavailable to do work (an increase in entropy). In addition, energy is constantly being passed from living organisms to nonliving matter (e.g., when you write you expend energy to move the pencil, etc.). The system of living organisms thus cannot be a static energy system, and must be replenished by energy derived from the nonliving world.

The second law of thermodynamics explains the loss of energy from the system at each successive trophic level in a food pyramid. In the food pyramid, the energy at the producer level is greater than the energy at the consumer I level which is greater than the energy of the consumer II level. Every energy transformation between the members of the successive levels involves the loss of usable energy, which increases entropy and causes the total amount of energy at each trophic level to be lower than at the preceding level.

1.5   Organic Chemistry

A) **Hydrocarbons** – The simplest organic molecules are the hydrocarbons. These compounds are composed solely of carbon and hydrogen. They can exist as chains (e.g., butane) or rings (e.g., benzene).

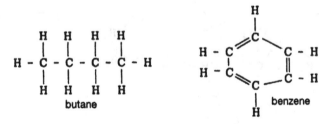

Figure 1.5 Butane and benzene chains

B) **Lipids** – Lipids are organic compounds that dissolve poorly, if at all, in water (hydrophobic). All lipids (fats and oils) are composed of carbon, hydrugcn, and oxygen where the ratio of hydrogen atoms to oxygen atoms is greater than 2:1. A lipid molecule is composed of 1 glycerol and 2 fatty acids.

Phospholipid – A phospholipid is a variety of a substituted lipid which contains a phosphate group.

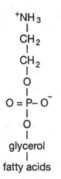

$$^+NH_3$$
$$|$$
$$CH_2$$
$$|$$
$$CH_2$$
$$|$$
$$O$$
$$|$$
$$O = P - O^-$$
$$|$$
$$O$$
$$|$$
glycerol
$$|$$
fatty acids

**Figure 1.6 An example of the addition which makes
a lipid a phospholipid**

C) **Steroids** – Steroids are complex molecules which contain carbon atoms arranged in four interlocking rings. Some steroids of biological importance are vitamin D, bile salts, and cholesterol.

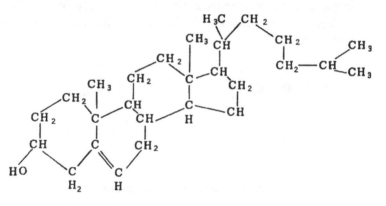

Figure 1.7 Structural formula of cholesterol

D) **Carbohydrates** – Carbohydrates are compounds composed of carbon, hydrogen, and oxygen, with the general molecular formula CH_2O. The principal carbohydrates include a variety of sugars.

1. **Monosaccharide** – a simple sugar or a carbohydrate which cannot be broken down into a simpler sugar. Its molecular formula is $(CH_2O)_n$ and the most common is glucose ($C_6H_{12}O_6$).

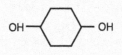

Figure 1.8 Glucose

2. **Disaccharide** - a double sugar, or a combination of two simple sugar molecules. Sucrose is a familiar disaccharide as are glucose and fructose.

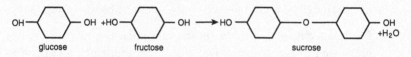

Figure 1.9 Double sugar formation by dehydration synthesis

3. **Polysaccharide** – a polysaccharide is a complex compound composed of a large number of glucose units. Examples of polysaccharides are starch, cellulose, and glycogen.

E) **Proteins** – All proteins are composed of carbon, hydrogen, oxygen, nitrogen, and sometimes phosphorus and sulfur. Approximately 50% of the dry weight of living matter is protein.

Amino Acids – The twenty amino acids are the building blocks of proteins.

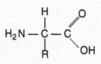

Figure 1.10 An amino acid with R representing its distinctive side chain

Polypeptides – Amino acids are assembled into polypeptides by means of peptide bonds. This is formed by a condensation reaction between the COOH groups and the NH_2 groups.

Primary Structure – The primary structure of protein molecules is the number of polypeptide chains and the number, type, and sequence of amino acids in each.

Secondary Structure – The secondary structure of protein molecules is characterized by the same bond angles repeated in successive amino acids which gives the linear molecule a recurrent structural pattern.

Tertiary Structure – The three-dimensional folding pattern, which is super-imposed on the secondary structure, is called the tertiary structure.

Quaternary Structure – The quaternary structure is the manner in which two or more independently folded subunits fit together.

F) **Nucleic Acids** – Nucleic acids are long polymers involved in heredity and in the manufacture of different kinds of proteins. The two most important nucleic acids are deoxyribonucleic acid (DNA) and ribonucleic acid (RNA).

Nucleotides – These are the building blocks of nucleic acids. Nucleotides are complex molecules composed of a nitrogenous base, a 5-carbon sugar, and a phosphate group.

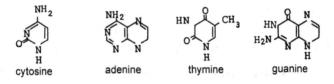

Figure 1.11 Structure of a nucleotide

Deoxyribonucleic Acid (DNA) – Chromosomes and genes are composed mainly of DNA. It is composed of deoxyribose, which is the ribose sugar missing the oxygen on carbon 2, nitrogenous bases, and phosphate groups.

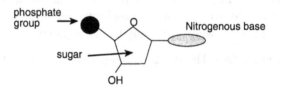

Figure 1.12 The four nitrogenous bases of DNA

Ribonucleic Acid (RNA) – RNA is involved in protein synthesis. Unlike DNA, it is composed of the sugar ribose and the nitrogenous base uracil instead of thymine.

Figure 1.13 Uracil

Problem Solving Examples:

Q Discuss some properties and functions of (a) carbohydrates, (b) lipids, (c) proteins, and (d) nucleic acids.

A (a) Carbohydrates are made up of carbon, oxygen and hydrogen, and have the general formula $(CH_2O)_n$. Carbohydrates can be classified as monosaccharides, disaccharides, oligosaccharides or polysaccharides. The monosaccharides ("simple sugars") are further categorized according to the number of carbons in the molecule. Trioses contain 3 carbons; pentoses contain 5 carbons (i.e., ribose, deoxyribose); and hexoses contain 6 carbons (i.e., glucose, fructose, galactose). The hexoses are important building blocks for complex carbohydrates.

Disaccharides, important in nutrition, are chemical combinations of two monosaccharides:

Lactose = glucose + galactose
Sucrose (table sugar) = glucose + fructose

Polysaccharides are complex carbohydrates made up of many monosaccharides. These long chains are formed by dehydration. They can also be broken down into monosaccharide units by hydrolysis. Many complex polysaccharides are of great biological significance whose primary functions include storage and structural properties. Examples of these are: starches (principal storage product of animals) and cellulose (major supporting material in plants).

(b) Lipids are also composed principally of carbon, hydrogen, and oxygen. However, they can also contain other elements, particularly phosphorus and nitrogen. Lipids are insoluble in water, but are soluble in different solvents. There are many known lipids; the most common are phospholipids and steroids. Fats are composed of two different types of compounds: glycerol (an alcohol) and fatty acids (organic compounds with a carboxyl group CO_2H). Each molecule of fat contains one glycerol molecule and three fatty acids joined together by dehydration reactions. There are basically two groups of fats: the saturated (those fats which have the maximum possible number of hydrogen atoms and therefore have no carbon to carbon double or triple bond) and the unsaturated (those which have at least one carbon to carbon double or triple bond). Unsaturated fats are an important part of our diet due to their storage capabilities. Phospholipids are composed of glycerol, fatty acids, phosphoric acid and a nitrogenous compound. They are important components of cellular membranes.

Steroids are classified as lipids because their solubility characteristics are similar to those of fats and phospholipids – insoluble in water, but soluble in ether. However, steroids differ structurally from fats and phospholipids. They are composed of four interlocking rings of carbon atoms with various side groups attached to the rings:

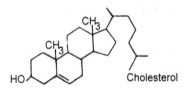

Cholesterol

They are very important biologically. Steroids include the sex hormones, various other hormones, and some vitamins. In addition, they are also important structural elements in living cells, especially plants.

(c) Proteins are much more complex than either carbohydrates or lipids. They are made up of the four essential elements: carbon, hydrogen, oxygen, and nitrogen. Some amino acids contain sulfur. Structurally, amino acids differ only by the "R" side chain, which can be very simple (as in glycine where R=H) or complex (as in tryptophan where R contains two ring structures). There are 20 amino acids which combine to form proteins. The bonds formed are called peptide bonds. A dipeptide is a molecule with two amino acids joined together by one peptide bond. Oligopeptides are short chains of amino acids. Polypeptides are polymers of amino acids. Finally, a protein is one or more polypeptide chains coiled or folded into complex three-dimensional configurations. Often a metal ion or organic molecule is part of the protein structure. Proteins are found in every part of the cell and are an integral part in both the structure and function of living things. Enzymes that catalyze chemical reactions in cells are proteins themselves. Proteins also function as the structural and binding materials of organisms. Hair, fingernails, muscle, cartilage, tendons, and ligaments are all structures which contain large amounts of proteins.

(d) Nucleic acids, as their name implies, are found primarily in the nucleus. There are two types of nucleic acid: deoxyribonucleic acid (DNA) and ribonucleic acid (RNA). DNA and RNA molecules are very long chains composed of repeating subunits called nucleotides. A nucleotide is composed of any one of the following five nitrogenous bases: adenine, guanine, cytosine, thymine (only in DNA) or uracil (only in RNA), a five carbon sugar (ribose in RNA, deoxyribose in DNA), and a phosphate group. The bases are attached to the sugars. Nucleic acids primarily function in heredity and governing the synthesis of many different kinds of substances present in organisms. Chromosomes and genes are predominantly composed of DNA. Some DNA is also found in the mitochondria and the chloroplasts.

 What is a polypeptide chain, and how is it related to the proteins in a cell?

A polypeptide chain consists of linked units called amino acids. Each amino acid consists of a carboxyl group (C terminal), an amino group (N-terminal), and a side chain which varies with each amino acid. An amino acid is represented as follows: amino acids can be linked together by peptide bonds, which form between the N terminal of an amino acid and a C terminal on another amino acid. When many amino acids are linked together by peptide bonds, we have a chain of amino acids known as a polypeptide chain.

Proteins are actually polypeptide chains. A protein molecule may be composed of one such chain, usually folded and cross-linked within itself to form a more compact and stable structure than a very long expanded chain. Some protein molecules are made of combinations of two or more polypeptide chains. The human protein hemoglobin, for instance, is composed of four polypeptides, two identical alpha chains, and two identical beta chains.

An important distinction must be made between polypeptides and proteins at the genetic level. A given gene can code for only one polypeptide. Many proteins, however, are composed of a number of polypeptides. When a protein is composed of more than one polypeptide chain, each chain may be coded for and made separately. These chains aggregate only after their individual syntheses to form the final protein. In addition, some proteins exist in their functional form only after certain changes have taken place in the original polypeptide. Thus, a given gene does not necessarily code directly for an active, functioning protein.

CHAPTER 2

Cellular Organization

2.1 The Cell Theory

The cell is the unit of structure and the unit of function of most living things. Cells arise from pre-existing cells by reproduction. The biochemical activities that occur within cells depend upon the presence of specific organelles.

Cell Membrane – The cell membrane is a double layer of lipids which surrounds a cell. Proteins are interspersed in this lipid bilayer. The membrane is semi-permeable; it is permeable to water but not to solutes.

The Nucleus – The nucleus is bounded by a pair of membranes. Within the nuclear membrane, there is a semi-fluid medium in which the chromosomes are suspended.

Cytoplasm – The cytoplasm is all the material in a cell located between the nucleus and the plasma membrane. Imbedded in the cytoplasm are the organelles.

Problem Solving Example:

Q What is a cell?

A A cell is the fundamental organizational unit of life. One of the most important generalizations of modern biology is the cell theory. There are two components of the cell theory. It states: (1) that all living things are composed of cells and (2) that all cells arise from other cells. Living things are chemical organizations of cells and capable of reproducing themselves. Functions of cells are due to specific and definite organelles.

There are many types of cells, and just as many classifications to go with them. There are plant cells, animal cells, eukaryotic cells, prokaryotic cells and many others. Also, within each of these divisions, there are smaller subdivisions pertaining to the specific properties or functions of the cells. Cells exhibit considerable variation in properties based on different arrangements of components. Cells also vary in size.

2.2 Cellular Organelles

A) **Mitochondria** – Mitochondria are organelles enclosed in a double-membrane. Mitochondria are the site of chemical reactions that extract energy from foodstuffs and make it available to the cell for all of its energy-demanding activities. This energy is provided in the form of ATP (adenosine tri-phosphate).

B) **Chloroplasts** – These are found only in the cells of plants and certain algae. Photosynthesis occurs in the chloroplasts.

C) **Plastids** – These structures are present only in the cytoplasm of plant cells. The most important plastid, chloroplast, contains chlorophyll, a green pigment.

D) **Lysosomes** – Lysosomes are membrane-enclosed bodies that function as storage vesicles for many digestive enzymes.

E) **Endoplasmic Reticulum (ER)** – The endoplasmic reticulum transports substances within the cell. There are two types: smooth and rough.

F) **Ribosomes** – These organelles are small particles composed chiefly of ribosomal-RNA and are the sites of protein synthesis. Rough ER has ribosomes attached.

G) **Golgi Apparatus** – The functions of the Golgi apparatus include storage, modification, and packaging of secretory products.

H) **Peroxisomes** – Peroxisomes are membrane-bound organelles which contain powerful oxidative enzymes.

I) **Vacuoles** – Vacuoles are membrane-enclosed, fluid-filled spaces. They have their greatest development in plant cells where they store materials such as soluble organic nitrogen compounds, sugars, various organic acids, some proteins, and several pigments.

J) **Cell Wall** – This is only present in plant cells and is used for protection and support.

K) **Centriole and Centrosome** – These function in cell division. They are present only in animals and protists (simple organisms such as bacteria, some algae, protozoa, and other organisms not easily classified as either plant or animal).

L) **Cilia and Flagella** – These are hairlike extensions from the cytoplasm of a cell. They both show coordinated beating movements, which are the major means of locomotion and ingestion in unicellular organisms.

M) **Nucleolus** – The nucleolus is a generally oval body composed of protein and RNA. Nucleoli are produced by chromosomes and participate in the process of protein synthesis.

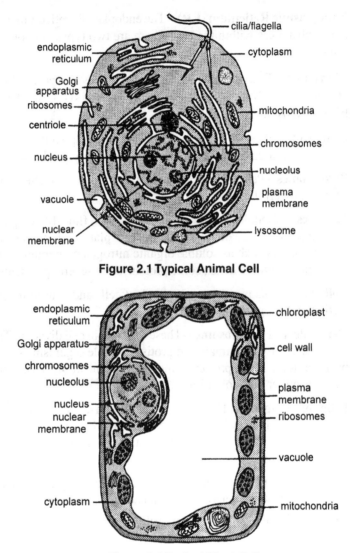

Figure 2.1 Typical Animal Cell

Figure 2.2 Typical Plant Cell

Problem Solving Examples:

 What is the structure and function of the chloroplasts in green plants?

 The chloroplasts have the ability to transform the energy of the sun into chemical energy stored in bonds that hold together the atoms of such foods (fuel) as glucose. By the process of photosynthesis, carbon dioxide and water react to produce carbohydrates with the simultaneous release of oxygen. Photosynthesis is driven forward by energy obtained from the sun.

Chlorophyll, the pigment contained in chloroplasts which gives plants their characteristic green color, is the molecule responsible for the initial trapping of light energy. Chlorophyll transforms light energy into chemical energy; then it passes this energy to a chain of other compounds involved in the energy-creating reactions.

Plastids are organelles which contain pigments and/or function in nutrient storage. Chloroplasts are but one example of a plastid; they give the green color to plants. The inner membrane gives rise to the complex internal system of the chloroplast. Surrounding the internal membranes is the stroma, which contains the enzymes which carry out the dark reactions of photosynthesis.

 In microscopy, small spherical bodies are often seen attached to the network of endoplasmic reticulum. What are these bodies? What function do they serve in the cell?

 The small spherical bodies that we see studding the endoplasmic reticulum (rER) – more accurately, the rough endoplasmic reticulum – are the ribosomes. The (rER) owes its rough appearance to the presence of ribosomes. The smooth endoplasmic reticulum appears smoother because it lacks ribosomes. Ribosomes consist of two parts, a large subunit and a small subunit. Both the large and small subunits are made of proteins and ribonucleic acid (RNA).

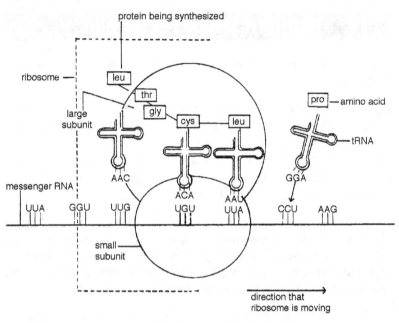

Diagram showing protein synthesis

Ribosomes are the sites of protein synthesis in the cell. Messenger RNA (mRNA), which carries genetic information from the nucleus to the ribosomes, associates with the small ribosomal subunit first and then binds to the large subunit as a prelude to protein synthesis. This association of mRNA to ribosomes makes the system of protein synthesis more efficient than if the complex was dispersed freely into the cytoplasm. The mRNA then pairs with complementary molecules of transfer RNA (tRNA), each carrying a specific amino acid, which bind with each other to form a highly specific protein molecule. Thus ribosomes are the sites where proteins are synthesized under genetic control.

 Why are the mitochondria referred to as the "powerhouse of the cell"?

 Mitochondria are membrane-bound organelles concerned principally with the generation of energy to support the various forms of chemical and mechanical work carried out by the cell. Mitochondria are distributed throughout the cell, because all parts of it require en-

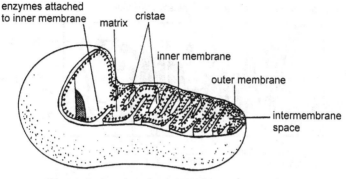

**Diagram showing the internal structures of a
mitochondrion through a cutaway view**

ergy. Mitochondria tend to be most numerous in regions of the cell that
consume large amounts of energy and more abundant in cells that require
a great deal of energy (for example, muscle and sperm cells).

Mitochondria are enclosed by two membranes. The outer one is a
continuous membrane. The inner membrane is thrown into many folds,
called cristae, that extend into the interior of the mitochondrion. Enclosed by the inner membrane is the matrix (see accompanying diagram), which contains many enzymes involved in the Krebs cycle.
Enzymes for the electron transport reactions are tightly bound to the
inner mitochondrial membrane.

The reactions that occur in the mitochondria all result in the production of ATP (adenosine triphosphate), which is the common cur-
rency of energy production in the cell. About 95 percent of all ATP
produced in the cell is created by the mitochondria. For this reason,
the mitochondria are commonly referred to as the powerhouse of the
cell.

 How is the Golgi apparatus related to the endoplasmic reticulum in function?

 The Golgi apparatus is composed of several membrane-
bounded, flattened sacs or cisternae (see the figure on the next
page). The sacs are disc-like and often slightly curved.

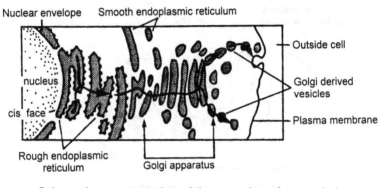

Nuclear envelope   Smooth endoplasmic reticulum

Outside cell

nucleus

Golgi derived
vesicles

cis face

Plasma membrane

Rough endoplasmic
reticulum

Golgi apparatus

**Schematic representation of the secretion of a protein in
a typical animal cell. The solid arrow represents the probable
route of secreted proteins.**

The function of the Golgi apparatus is best understood in cells involved in protein synthesis and secretion. The protein to be secreted is synthesized on the rough endoplasmic reticulum. Vesicles containing small quantities of the synthesized protein bud off from the endoplasmic reticulum. These vesicles carry the protein to the Golgi complex. In the Golgi apparatus, the protein is concentrated by the removal of water. In addition, chemical modifications of the protein, such as glycosylation (addition of sugar) occur. The modified protein is released in the form of secretory granules. The secretory granules containing the protein can fuse with the plasma membrane and expel its contents from the cell, a process known as exocytosis.

Most of the cell organelles are found in a specific arrangement within the cell to complement their function. For example, the Golgi apparatus is usually found near the cell membrane and associated with the endoplasmic reticulum. Since they are relatively close to each other, transport of materials between them is considerably efficient.

2.3 Prokaryotes vs. Eukaryotes

Prokaryote – Prokaryote refers to bacteria and blue-green algae. Prokaryotic cells have no nuclear membrane, and they lack membrane-bound subcellular organelles such as mitochondria and chloroplasts.

However, the membrane that bounds the cell is folded inward at various points, and carries out many of the enzymatic functions of many internal membranes of eukaryotes.

Eukaryote – Eukaryote refers to all the protists, plants, and animals. These are characterized by true nuclei; bounded by a nuclear membrane, and containing membrane-bound subcellular organelles.

Table 2.1 A comparison of eukaryotic and prokaryotic cells

Characteristic	Eukaryotic cells	Prokaryotic Cells
Chromosomes	multiple, composed of nucleic acids and protein	single, composed only of nucleic acid
Nuclear Membrane	present	absent
Mitochondria	present	absent
Golgi apparatus, endoplasmic reticulum, lysosomes, peroxisomes	present	absent
Photosynthetic apparatus	chlorophyll, when present is contained in chloroplasts	may contain chlorophyll
Microtubules	present	rarely present
Ribosomes	large	small
Flagella	have 9-2 tubular structure	lack 9-2 tubular structure
Cell wall	when present, does not contain muramic acid	contains muramic acid

Problem Solving Examples:

 Even though there is no such thing as a "typical cell" – for there are too many diverse kinds of cells – biologists have determined that there are two basic cell types. What are these two types of cells?

Cells are classified as either prokaryotic or eukaryotic. A key difference between the two cell types is that prokaryotic cells lack the nuclear membrane characteristic of eukaryotic cells. Prokaryotic cells have a nuclear region, which consists of nucleic acids. Eukaryotic cells have a nucleus, bound by a double-layered membrane. The eukaryotic nucleus consists of DNA which is bound to proteins and organized into chromosomes.

Bacteria and blue-green algae are prokaryotic unicellular organisms. Other organisms, for example, protozoa, algae, fungi, higher plants, and animals are eukaryotic. Eukaryotic cells have discrete regions that are usually delimited from the rest of the cell by membranes. These are called membrane-bound subcellular organelles. They perform specific cellular functions, for example, respiration and photosynthesis. The enzymes for these processes are located within membrane-bound mitochondria and chloroplasts, respectively. In prokaryotic cells, there are no such membrane-bound organelles. Respiratory and photosynthetic enzymes are not segregated into discrete organelles although they have an orderly arrangement. Prokaryotic cells lack endoplasmic reticulum, Golgi apparatus, lysosomes, and vacuoles. In short, prokaryotic cells lack the internal membranous structure characteristic of eucaryotic cells.

There are other differences between prokaryotic cells and eukaryotic cells. The flagella of bacteria are structurally different from eukaryotic flagella.

What are the chief components and structures of a eukaryotic cell?

Membranes, composed of lipids and proteins, are a crucial component of both animal and plant cells. The plasma membrane surrounds the cell and serves to separate the internal living matter from the external environment. Plant cells have a cell wall external to the plasma membrane. It is composed mainly of cellulose and is fairly rigid but is permeable. Selectivity of materials entering the plant cells is a function of the plasma membrane. Membranes inside the cell divide the cell into compartments distinctive in both form and function. Subcellular structures in the cytoplasm are known as organelles. When these are surrounded by membranes, they are termed membrane-bound organelles. Each organelle performs some specific functions.

The nucleus is the controlling center of the cell. The nucleus is surrounded by two layers of membrane which form the nuclear envelope. Within the nucleus are found the chromosomes, composed chiefly of deoxyribonucleic acid (DNA) and protein. Genes located within the chromosomes direct cellular function and are capable of being replicated in nuclear division. The nucleolus is a specialized region in the nucleus involved in the synthesis of ribosomal RNA, the material making up the ribosomes.

Mitochondria and chloroplasts are membrane-bound organelles involved in energy production in the cell. Chloroplasts in plant cells convert solar energy to chemical energy contained in organic substances, which are oxidized in the mitochondria to yield energy in the form of ATP.

Cells which utilize food for energy must take food into the cell and degrade it. Lysosomes are membrane-bound organelles whose digestive enzymes break down organic substances to simpler forms, which can be used by the cell to yield energy.

Most cells contain membrane-bound bodies called vacuoles. Small vacuoles may be termed vesicles. These structures may contain the ingested materials taken from the cell exterior or materials to be released by the cell. Mature plant cells usually contain a single large fluid-filled vacuole, which aids the cell in maintaining an internal pressure and rigidity.

Organelles involved in the synthesis and transport of cellular components are the ribosomes, endoplasmic reticulum, and Golgi apparatus. Ribosomes are involved in protein synthesis. The endoplasmic reticulum is a system of membranes providing transport channels within the cell. It transports the synthesized protein to other parts of the cell. The Golgi apparatus is involved in the packaging of cellular products before they are released outside of the cell.

Organelles involved in the maintenance of cellular shape and in movement are the microfilaments and microtubules, also known as the skeleton of the living cell. Microfilaments are involved in the connection of adjacent cells for intercellular communication. They also function in the transport of products within the cell. The microtubules are the basic substance in the cilia and flagella of motile cells, which move by the contracting action of microtubules. The microtubules have also been identified as the fundamental substance in the spindle apparatus during cell division. Both microfilaments and microtubules are protein structures.

2.4 Tissues

A tissue is a group of similarly specialized cells which together perform certain special functions.

2.4.1 Plant Tissues

A mature vascular plant possesses several distinct cell types which group together in tissues. The major plant tissues include epidermal, parenchyma, sclerenchyma, chlorenchyma, vascular, and meristematic.

Table 2.2 Summary of Plant Tissues

Tissue	Location	Functions
1. Epidermal	Root	Protection-Increases absorption area
	Stem	Protection-Reduces H_2O loss
	Leaf	Protection-Reduces H_2O loss Regulates gas exchange
2. Parenchyma	Root, stem, leaf	Storage of food and H_2O
3. Sclerenchyma	Stem and leaf	Support
4. Chlorenchyma	Leaf and young stems	Photosynthesis
5. Vascular		
a. Xylem	Root, stem and leaf	Upward transport of fluid
b. Phloem	Leaf, root, stem	Downward transport of fluid
6. Meristematic	Root and stem	Growth; formation of xylem, phloem, and other tissues

2.4.2 Animal Tissues

The cells that make up multicellular organisms become differentiated in many ways. One or more types of differentiated cells are organized into tissues. The basic tissues of a complex animal are the epithelial, connective, nerve, muscle, and blood tissues.

Table 2.3 Summary of Animal Tissues

Tissue	Location	Functions
1. Epithelial	Covering of body	Protection
	Lining internal organs	Secretion
2. Muscle		
a. Skeletal	Attached to skeleton bones	Voluntary movement

b. Smooth	Walls of internal organs	Involuntary movement
c. Cardiac	Walls of heart	Pumping blood
3. Connective		
a. Binding	Covering organs, in tendons and ligaments	Holding tissues and organs together
b. Bone	Skeleton	Support, protection, movement
c. Adipose	Beneath skin and around internal organs	Fat storage, insulation, cushion
d. Cartilage	Ends of bone, part of nose and ears	Reduction of friction, support
4. Nerve	Brain	Interpretation of impulses, mental activity
	Spinal cord, nerves, ganglions	Carrying impulses to and from all organs
5. Blood	Blood vessels, heart	Carrying materials to and from cells, carrying oxygen, fighting germs, clotting

Problem Solving Examples:

Q List and compare the tissues that support and hold together the other tissues of the body.

A Connective tissue functions to support and hold together structures of the body. They are classified into four groups by structure and/or function: bone, cartilage, blood, and fibrous connective tissue. The cells of these tissues characteristically secrete a large amount of noncellular material, called matrix. The nature and function of each kind of connective tissue is determined largely by the nature of its matrix. Connective tissue cells are actually quite separate from each other, for most of the connective tissue volume is made up of matrix. The cells between them function indirectly, by secreting a matrix which performs the actual functions of connection or support or both.

Blood consists of red blood cells, white blood cells, and platelets in a liquid matrix called the plasma. Blood has its major function in transporting almost any substance that is needed, anywhere in the body.

The fibrous connective tissues are composed of interlacing protein fibers secreted by and surrounding the connective tissue cells. These fibers are of three types: collagenous, elastic, and reticular. These fibrous tissues occur throughout the body and hold skin to muscle, keep glands in position, and bind together many other structures. Tendons and ligaments are specialized types of fibrous connective tissue. Tendons are not elastic but are flexible, cable-like cords that connect muscles to each other or to bones. Ligaments are semi-elastic and connect bones to bones.

The supporting skeleton of vertebrates is composed of the connective tissues cartilage and bone. Cartilage cells secrete a hard rubbery matrix around themselves. Cartilage can support great weight, yet it is flexible and somewhat elastic. Cartilage is found in the human body at the tip of the nose, in the ear flaps, the larynx and trachea, intervertebral discs, surfaces of skeletal joints and ends of ribs, to name a few places.

Bone has a hard, relatively rigid matrix. This matrix contains many collagenous fibers and water, both of which prevent the bone from being overly brittle. Bone is impregnated with calcium and phosphorus salts. These give bone its hardness. Bone is not a solid structure, for most bones have a large marrow cavity in their centers.

 Describe the composition and structure of the plant cell wall.

Because plant cells must be able to withstand high osmotic pressure differences, they require rigid cell walls to prevent bursting. This rigidity is provided primarily by cellulose, the most abundant cell wall component in plants. Cellulose, a polysaccharide, is made of glucose molecules. In the cell wall, cellulose molecules are organized in bundles of parallel chains to form fibrils. The fibrils are often arranged in reinforced, criss-crossed layers, providing a matrix capable of withstanding enormous stress.

A layer known as the middle lamella lies between and is shared by adjacent cells. By binding the cells together, it provides additional stiffness to the plant.

What are the chief differences between plant and animal cells?

 A study of both plant and animal cells reveals the fact that in their most basic features, they are alike. However, they differ in several important ways. First of all, plant cells, but not animal cells, are surrounded by a rigid cellulose wall. The cell wall is actually a secretion from the plant cell. It surrounds the plasma membrane and is responsible for the maintenance of cell shape. Animal cells, without a cell wall, cannot maintain a rigid shape.

Most mature plant cells possess a single large central fluid sack, the vacuole. Vacuoles in animal cells are small and frequently numerous.

Another distinction between plant and animal cells is that many of the cells of green plants contain chloroplasts, which are not found in animal cells. The presence of chloroplasts in plant cells enables green plants to be autotrophs, organisms which synthesize their own food. Animal cells, devoid of chloroplasts, cannot produce their own food. Animals, therefore, are heterotrophs, organisms that depend on other living things for nutrients.

2.5 Exchange of Materials Between Cell and Environment

Diffusion - The net migration of molecules or ions as a result of their own random movements from a region of higher concentration to a region of lower concentration is known as diffusion.

Osmosis - Osmosis is the net movement of water through a semipermeable membrane. At constant temperature and pressure, the net movement of water is from the solution with lower concentration to the solution with higher concentration of osmotically active particles.

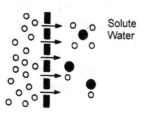

Solute
Water

Figure 2.3 The process of osmosis

Active Transport – The movement of ions and molecules against a concentration gradient is referred to as active transport. The cell must expend energy to accomplish the transport. In passive transport, no energy is expended.

Endocytosis – Endocytosis is an active process in which the cell encloses a particle in a membrane-bound vesicle, pinched off from the cell membrane. Endocytosis of solid particles is called phagocytosis.

prey

food vacuole

Figure 2.4 Endocytosis in the amoeba

Exocytosis – Exocytosis is the reverse of endocytosis; there is a discharge of vacuole-enclosed materials from a cell by the fusion of the cell membrane with the vacuole membrane.

Isotonic Medium – An isotonic medium is one in which the cell is in osmotic balance because it contains the same concentration of osmotically active particles.

Hypertonic Medium – A hypertonic medium is one in which a cell loses water because the medium contains a higher concentration of osmotically active particles.

Hypotonic Medium – A hypotonic medium is one in which a cell gains water because the medium contains a lower concentration of osmotically active particles.

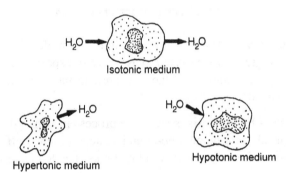

Figure 2.5 Osmotic effects of the fluids bathing cells

Problem Solving Examples:

 Differentiate clearly between diffusion and osmosis.

 Diffusion is the general term for the net movement of the particles of a substance from a region where the substance is at a high concentration to regions where the substance is at a low concentration. When the movements of all the particles are considered jointly, there is a net movement away from the region of high concentration toward regions of low concentration. This results in the particles of a given substance distributing themselves with relatively uniform density or concentration within any available space. Diffusion tends to be faster in gases than in liquids, and much slower in solids.

The movement or diffusion of water or solvent molecules through a semipermeable membrane is called osmosis. The osmotic pressure of a solution is a measure of the tendency of the solvent to enter the solution by osmosis. The more concentrated a solution, the more water will tend to move into it, and the higher is its osmotic pressure.

 Why is the phenomenon of diffusion important to movement of materials in living cells?

In a living cell, chemical reactions are constantly taking place to produce the energy or organic compounds needed to maintain life. The reacting materials of chemical reactions must be supplied continuously to the actively metabolizing cell, and the products distributed to other parts of the cell where they are needed. This is extremely important because if the reactants are not supplied, the reaction ceases. If the products are not distributed but instead accumulate near the site of reaction, Le Châtelier's principle of chemical reactions operates, driving the reversible reaction backward and diminishing the concentration of the products. Thus, in order to maintain a constant chemical reaction, the reactants must be continuously supplied and the products must move through the cell medium to other sites. Diffusion is how the products are distributed.

When a certain chemical reaction is operating in the cell, some reacting substance will be consumed. The concentration of this substance is lower in regions closer to the site of reaction than regions farther away from it, establishing a concentration gradient. The concentration gradient causes the reactant to move from a region of higher concentration to a region of lower concentration, or the reaction site. This movement is called diffusion. Thus, by diffusion, molecules tend to move to regions in the cell where they are being consumed. The products of the reaction travel away from the reaction site also by this process. At the reaction site, the concentration of the products is highest, hence the products tend to move away from this region to ones where they are lower in concentration. The removal of products signals the reaction to keep on going. When the product concentration gets too high, the reaction is inhibited by a built-in feedback mechanism.

Thus, diffusion explains the occurrence of movement of chemical substances into or out of the cell and within the cell. For example, oxygen molecules follow a concentration gradient to enter the cell and move toward the mitochondria, because oxygen concentration is the lowest in the mitochondria where oxidation reactions continually consume oxygen. Carbon dioxide is produced in the mitochondria when an acetyl unit is completely oxidized in the citric acid cycle. The CO_2 will then travel away from the mitochondria to other parts of the cell or out of the cell into the bloodstream where it is lower in concentration.

Quiz: The Basis of Life – Cellular Organization

1. Prokaryotic cells differ from eukaryotic cells in that the former lack

    (A) ribosomes.

    (B) a plasma membrane.

    (C) endoplasmic reticulum.

    (D) a cell wall.

2. Unlike plant cells, animal cells possess

    (A) a cell wall.

    (B) centrioles.

    (C) chloroplasts.

    (D) a nuclear membrane.

3. Sperm cells are highly motile cells and require a great deal of energy to maintain their activity. An organelle that would be found in great abundance in this cell is the

    (A) mitochondrion.

    (B) ribosome.

    (C) lysosome.

    (D) testosterone.

4.   The Golgi apparatus primarily functions in

    (A)   packaging protein for secretion.

    (B)   synthesizing protein for secretion.

    (C)   packaging protein for hydrolysis.

    (D)   both (A) and (B).

5.   Unlike chlorenchyma and sclerenchyma tissues, parenchyma tissue does not function in

    (A)   support.

    (B)   gas exchange.

    (C)   nutrient exchange.

    (D)   both (B) and (C).

6.   A normal human erythrocyte contains 0.154 molar of sodium salts; the sodium ion is a non-penetrating particle to the erythrocyte membrane. When placed in a solution of 0.2 molar sodium chloride, the erythrocyte will

    (A)   hemolyze (burst) due to an influx of water.

    (B)   hemolyze due to an influx of sodium.

    (C)   crenate (shrink) due to an efflux of water.

    (D)   crenate due to an efflux of sodium.

7. The function of phloem is to

   (A) cover and protect.

   (B) convert nutrients from the soil.

   (C) strengthen and support.

   (D) transport organic solutes.

8. A sudden shutdown of a cell's ribosomes would result in a build-up of which of the following molecules?

   (A) Carbon dioxide

   (B) Carbon monoxide

   (C) Lysine

   (D) Oxygen

9. All living organisms are classified as eukaryotes (true nucleus) or prokaryotes (before the nucleus). An example of a prokaryote is

   (A) the AIDS virus.

   (B) *E. coli.*

   (C) *Homo Sapiens.*

   (D) an oak tree.

10. What is the fundamental difference between active and passive transport across cell membranes?

    (A) Active transport occurs more rapidly than passive transport.

    (B) Passive transport is never selective.

    (C) Passive transport requires a concentration gradient across the cell membrane as the driving force, while active transport needs energy expenditure to transport substances regardless of concentration gradient.

    (D) Passive transport occurs only among gases.

ANSWER KEY

1.	(C)	6.	(C)
2.	(B)	7.	(D)
3.	(A)	8.	(C)
4.	(A)	9.	(B)
5.	(A)	10.	(C)

CHAPTER 3

Cellular Metabolism and Energy Pathways

3.1 Photosynthesis

Photosynthesis is the basic food-making process through which inorganic CO_2 and H_2O are transformed into organic compounds, specifically carbohydrates.

Chloroplasts absorb light energy and use CO_2 and H_2O to synthesize carbohydrates. Oxygen, which is formed as a by-product, is either eliminated into the air through the stomata, stored temporarily in the air spaces, or used in cellular respiration.

An overall chemical description of photosynthesis is the equation

$$6\,CO_2 + 6\,H_2O \xrightarrow[\text{chlorophyll}]{\text{light}} C_6H_{12}O_6 + 6\,O_2$$

3.1.1 Light Reaction (Photolysis)

A first step in photosynthesis is the decomposition of water molecules to separate hydrogen and oxygen components. This decomposition is associated with processes involving chlorophyll and light and is thus known as the light reaction.

3.1.2 Dark Reaction (CO_2 Fixation)

In this second phase, the hydrogen that results from photolysis reacts with CO_2, and carbohydrate forms. CO_2 fixation does not require light.

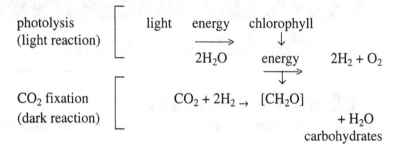

Figure 3.1 Photolysis and CO_2 fixation

Problem Solving Example:

 Discuss the sequence of reactions that constitute the "dark reactions" of photosynthesis. What are the products of these reactions?

 The dark reactions of photosynthesis, in which carbohydrates are synthesized, occur in a cyclic sequence of three phases – the carboxylative, reductive, and regenerative phases. These dark reactions do not require the presence of sunlight.

In the carboxylative phase, a five-carbon sugar, ribulose-5-phosphate, is phosphorylated by ATP to yield ribulose diphospate, which is then carboxylated (that is, CO_2 is added), yielding a six-carbon intermediate which is split immediately by the addition of water. This carbohydrate form is reduced by NADPH in an enzymatic reaction in the reductive phase, utilizing energy from ATP. The product of this reduction is a triose. Two molecules of the triose condense in the regenerative phase to form one hexose molecule, known as fructose. This is subsequently converted to a glucose-6-phosphate molecule and finally transformed into starch.

The materials consumed in the production of one hexose molecule are one molecule each of CO_2 and H_2O, three ATP molecules, and four H atoms (from two molecules of $NADPH_2$). The final product can either be polymerized into starch and stored or broken down to yield energy for work.

3.2 Cellular Respiration

During cellular respiration, glucose must first be activated before it can break down and release energy. After activation, glucose enters into numerous reactions occurring in two stages. One stage is the anaerobic phase of cellular respiration, and the other is the aerobic phase.

A) **Glycolysis** – Glycolysis refers to the breakdown of glucose which marks the start of the anaerobic reactions of cellular respiration. ATP is the energy source which activates glucose and initiates the process of glycolysis.

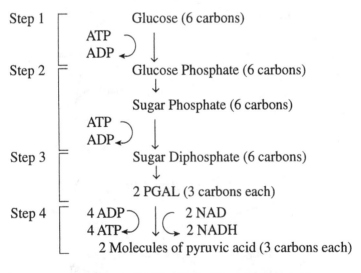

Step 1   Glucose (6 carbons)

      ATP
      ADP

Step 2   Glucose Phosphate (6 carbons)

      Sugar Phosphate (6 carbons)

      ATP
      ADP

Step 3   Sugar Diphosphate (6 carbons)

      2 PGAL (3 carbons each)

Step 4   4 ADP   2 NAD
          4 ATP   2 NADH
          2 Molecules of pyruvic acid (3 carbons each)

Figure 3.2 The Major Steps in Glycolysis

The steps in Figure 3.2 are summarized as follows:

Step 1 – Activation of glucose

Step 2 – Formation of sugar diphosphate

Step 3 – Formation and oxidation of PGAL, phosphoglyceraldehyde

Step 4 – Formation of pyruvic acid ($C_3H_4O_3$)
      Net gain of two ATP molecules

B) **Krebs Cycle (Citric Acid Cycle)** – The Krebs cycle is the final common pathway by which the carbon chains of amino acids, fatty acids, and carbohydrates are metabolized to yield CO_2. Pyruvic acid is converted to acetyl coenzyme A and, through a series of reactions, citric acid is formed.

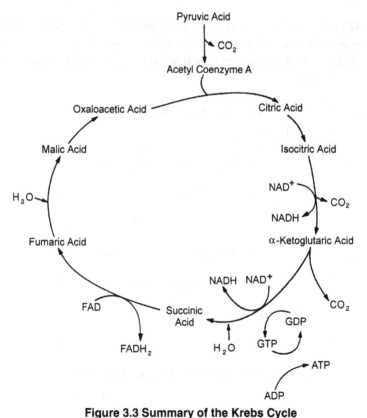

Figure 3.3 Summary of the Krebs Cycle

Problem Solving Example:

 Explain the events which take place during glycolysis.

 Glycolysis is the series of metabolic reactions by which glucose is converted to pyruvate with the concurrent formation of ATP. Glycolysis occurs in the cytoplasm of the cell, and the presence of oxygen is unnecessary. Glucose is first phosphorylated by a high-energy phosphate from ATP. The product, glucose-6-phosphate, undergoes rearrangement to fructose-6-phosphate, which is subsequently phosphorylated by another ATP to yield a sugar phosphate (6 carbons). This hexose is then split into two three-carbon sugars, glyceraldehyde-3-phosphate (PGAL). Only PGAL and another product can be directly degraded in glycolysis that is reversibly converted into PGAL by enzyme action.

Since two molecules of PGAL are thus produced per molecule of glucose oxidized, the products of the subsequent reactions can be considered "doubled" in amount.

$$1 \text{ glucose} \longrightarrow 2 \text{ PGAL}$$

PGAL gets oxidized and phosphorylated. NAD^+ is reduced to NADH, and the product of this reaction reacts with ADP to form ATP and a phosphated triose. After a dehydrogenating reaction, phosphoenolpyruvate (PEP) is formed. Finally, this phosphate group is transferred to ADP, yielding ATP and pyruvate.

Since two molecules of PGAL are formed per molecule of glucose, four ATP molecules are produced during glycolysis. The net yield of ATP is only 2, since 2 ATP were utilized in initiating glycolysis (reactions 1 and 3). Pyruvate is then converted to acetyl coenzyme A which enters the Krebs cycle. In addition, two molecules of NADH are produced per molecule of glucose. Hence, the net result of glycolysis is that glucose is degraded to pyruvate with the net formation of 2 ATP and 2 NADH. The process of glycolysis can be summarized as follows:

$$\text{glucose} + 2ADP + 2P_i + 2NAD^+ \longrightarrow$$
$$2 \text{ pyruvate} + 2ATP + 2NADH + 2H^+$$

3.3 ATP and NAD

ATP – ATP stands for adenosine triphosphate and is a coenzyme essential for the breakdown of glucose. When the bonds in ATP are hydrolyzed, a large amount of energy is released. It takes 7 kcal to phosphorylate (add a phosphate) to ADP to make ATP.

NAD – NAD stands for nicotinamide adenine dinucleotide. Like ATP, it is also a coenzyme. NAD participates in a large number of oxidation-reduction reactions in cells, including those in cellular respiration. It is an electron acceptor and donor. Oxidation of NADH releases 53 kcal.

Problem Solving Examples:

 NAD$^+$ and NADP$^+$ play important roles in reactions associated with metabolism in both plants and animals. Discuss these roles.

 In a dehydrogenation reaction, the electrons removed from a substance cannot exist freely but must be transferred to another compound called an electron acceptor. Both nicotinamide adenine dinucleotide (NAD$^+$) and nicotinamide adenine dinucleotide phosphate (NADP$^+$) are electron acceptors. The functional part of both NAD$^+$ and NADP$^+$ is the nicotinamide ring, which accepts one hydrogen atom and two electrons.

NAD$^+$ is used as an electron acceptor in glycolysis, the Krebs (or citric acid [TCA]) cycle, and fatty acid oxidation. NADP$^+$ is used as an electron acceptor in the light reaction of photosynthesis. We thus see that the function of NAD$^+$ or NADP$^+$ is to accept hydrogen atoms and electrons and to store energy released from oxidized substances in the form of NADH or NADPH. The reverse is true for NADH or NADPH; that is, their function is to give up hydrogen atoms and electrons and energy to reducible compounds. For example, the hydrogen atoms of NADH in the electron transport chain are ultimately transferred to oxygen, producing high energy phosphate bonds in the form of ATP along the chain. The oxidation of one molecule of NADH yields 3 phosphate bonds; thus the 8 NADHs produced in the citric acid cycle yield 24 phosphate bonds.

NADPH is formed from the reduction of $NADP^+$ in the light reaction of photosynthesis. It is subsequently utilized in the dark reactions of carbohydrate synthesis in which carbon dioxide is used in the production of hexose molecules. To produce one hexose molecule (a 6-carbon sugar), 12 NADPH and 18 ATP are required, both of which are the products of the light reactions. Thus, NAD^+, NADH and $NADP^+$, NADPH have major importance in reactions associated with metabolic and catabolic functions.

 When is ATP generated? What is the advantage of having a sequential transfer of electrons rather than one single transfer?

 The NADH and $FADH_2$ formed in glycolysis and the citric acid (TCA) cycle are energy-rich molecules. When each of these molecules transfers its pair of electrons to molecular oxygen, a large amount of energy is released. This released energy can be used to generate ATP in oxidative phosphorylation. As electrons flow through the electron transport chain from NADH to O_2, ATP is formed at three sites along the chain.

The advantage of the sequential transfer over one single transfer is that the sequential reactions divide up the free energy change of the oxidation of NADH, which is highly exergonic. If all this energy were released at once, much of it would be wasted as heat since only 7 kcal are needed to phosphorylate ADP to form ATP. If all the energy from the oxidation of NADH is released at once to form 1 ATP, the efficiency of this system is

$$^7/_{53} \times 100\% = 13\%$$

However, if the same energy is released in a stepwise process to form 3 ATP, the efficiency is

$$^{21}/_{53} \times 100\% = 40\%$$

These stepwise reactions increase the efficiency of oxidative phosphorylation from 13% to 40%.

In addition, the large amount of heat liberated would very easily destroy enzyme activity and thus is harmful to the cell if released in one burst.

The three reactions involved are "coupled" with the phosphorylation of ADP so that the liberated energy is used immediately to drive this otherwise non-spontaneous reaction.

The exact mechanism of oxidative phosphorylation, that is, how the energy liberated is transferred so as to form the high energy phosphate bond of ATP, is unknown.

3.4 The Respiratory Chain (Electron Transport System [ETS])

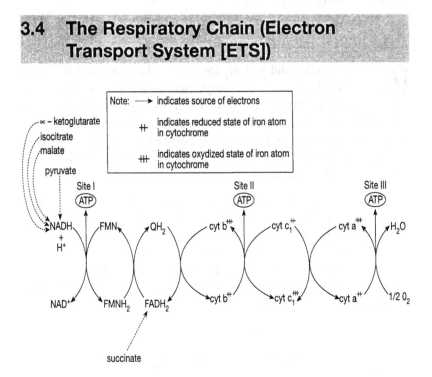

Figure 3.4 The Respiratory Chain

This system consists of a series of enzymes and coenzymes which pick up, hold, and then transfer hydrogen atoms among themselves until the hydrogen reaches its final acceptor, which is oxygen. Cytochromes are the enzymes and coenzymes involved in transferring hydrogen.

The cytochromes, together with other enzymes, split hydrogen atoms attached to compounds such as $NADH_2$ into hydrogen ions and electrons. Each cytochrome then passes the hydrogen ions and electrons to another cytochrome in the series.

Problem Solving Example:

Q Describe the pathway of electron transport in the respiratory chain.

A A chain of electron carriers is responsible for transferring electrons through a sequence of steps from molecules such as NADH to molecular oxygen. NADH collects electrons from many different substrates through dehydrogenases. For example, NADH is produced by oxidation in glycolysis, and by several dehydrogenations in the citric acid cycle. In the electron transport system, NADH is oxidized to NAD^+ with the corresponding reduction of flavin mononucleotide (FMN) to $FMNH_2$, producing one ATP molecule. Then $FMNH_2$ is oxidized back to FMN, and the electrons of $FMNH_2$ are transferred to ubiquinone (coenzyme Q). It is at this site that electrons of $FADH_2$ are funneled into the electron transport chain.

As ubiquinone is oxidized, the first of a series of cytochromes, cytochrome b, is reduced. The cytochromes are large heme proteins. A heme group consists of an iron atom surrounded by a flat organic molecule called a porphyrin ring.

Cytochrome b reduces cytochrome c; the energy released is used for the formation of a second ATP molecule, which in turn reduces cytochrome a. Molecular oxygen is the final electron acceptor. As oxygen is reduced to H_2O, a third and last ATP molecule is synthesized.

Thus for each NADH molecule entering the respiratory chain, 3

ATP molecules are produced. Therefore, eight NADH molecules give rise to 24 ATP per glucose molecule in the citric acid cycle. Only 4 ATP are produced by the 2 $FADH_2$ molecules, since they enter the electron transport system at ubiquinone, thus bypassing the first site of ATP synthesis.

The two cytoplasmic NADH from glycolysis produce only four ATP instead of six because one ATP is expended per cytoplasmic NADH in order to actively transport NADH across the mitochondrial membrane. By means of the respiratory chain, 32 ATP are produced per glucose molecule. The other 4 ATP molecules are produced during substrate-level phosphorylation in glycolysis and the TCA cycle. Thus there is a net yield of 36 ATP per glucose molecule oxidized. (Recall that "substrate-level" indicates reactions not involving the electron transport system.)

3.5   Enzymes

Enzymes are protein catalysts that lower the amount of activation energy needed for a reaction, allowing it to occur more rapidly. The enzyme binds with the substrate but resumes its original conformation after completion of the reaction.

Substrate – A substrate is the molecule upon which an enzyme acts. The atoms of the substrate are arranged so as to fit into the active site of the enzyme that acts upon it. An enzyme usually affects only one substrate.

SUBSTRATE

PRODUCT

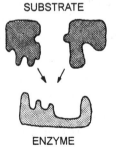

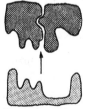

ENZYME

Enzyme-substrate
complex

Enzyme resumes
original conformation

Figure 3.5 The action of an enzyme

Allosteric Enzyme – An allosteric enzyme is one that can exist in two distinct conformations. Usually, the enzyme is active in one conformation and inactive in the other.

Allosteric Inhibition – During the process of allosteric inhibition, an inhibitory molecule called a negative modulator binds to the enzyme and stabilizes it in its inactive conformation, preventing the reaction from proceeding.

Coenzymes – These are metal ions or non-proteinaceous organic molecules that bind briefly and loosely to some enzymes. The coenzyme is necessary for the catalytic reaction of such enzymes.

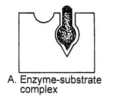

A. Enzyme-substrate complex   B. Competitive inhibitor bound to enzyme   C. Noncompetitive inhibitor bound to enzyme

Figure 3.6 A. Enzyme-substrate complex; B. Competitive inhibition; C. Non-competitive inhibition

Factors influencing the rate of enzyme action:

1. pH

2. temperature

3. concentration of enzyme and substrate

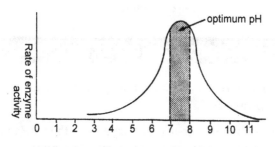

Figure 3.7 Effect of pH on rate of enzyme action

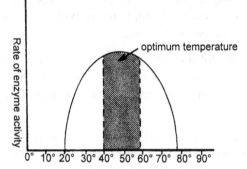

Figure 3.8 Effect of temperature upon rate of enzyme activity

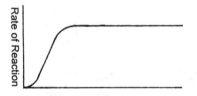

Figure 3.9 Fixed amount of enzyme and an excess
of substrate molecules

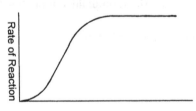

Figure 3.10 Fixed number of substrate molecules and an excess of
enzyme molecules

Problem Solving Examples:

 What are some of the important properties and characteristics
of enzymes?

 An important property of enzymes is their catalytic ability.
Enzymes control the speed of many chemical reactions that
occur in the cell. To understand the efficiency of an enzyme, one can

measure the rate at which an enzyme operates – also called the turn-over number. The turnover number is the number of molecules of sub-strate which is acted upon by a molecule of enzyme per second. Most enzymes have high turnover numbers and are thus needed in the cell in relatively small amounts.

A second important property of enzymes is their specificity where an enzyme is usually configured to "fit" particular substrates. Each enzyme has a region called a binding site to which only certain sub-strate molecules can bind efficiently. There are varying degrees of specificity: urease, which decomposes urea to ammonia and carbon dioxide, will react with no other substance; however, lipase will hy-drolyze the ester bonds of a wide variety of fats.

The structures of different enzymes differ significantly. Some are composed solely of protein (for example, pepsin). Others consist of two parts: a protein part (also called apoenzyme) and a non-protein part, either an organic coenzyme or an inorganic cofactor, such as a metal ion. Only when both parts are combined can activity occur.

There are other important considerations. Enzymes, as catalysts, do not determine the direction of a reaction, but they affect the rate at which the reaction reaches equilibrium. Enzymes are efficient because they are needed in very little amounts and can be used repeatedly. As enzymes are proteins, they can be permanently inactivated or dena-tured by extremes in temperature and/or pH, and also have an opti-mal temperature or pH range within which they work most efficiently.

 Describe the mode of action of enzymes. What factors affect enzyme activity?

 An enzyme (E) combines with its substrate (S) to form an in-termediate enzyme-substrate complex (ES), which then de-composes into reaction products (P) and the free enzyme, as seen in the following equation.

$$E + S \longrightarrow ES \longrightarrow E + P$$
$$\text{enzyme-substrate}$$
$$\text{complex}$$

An enzyme causes the substrate upon which it is acting to be much more reactive than when it is free. One postulate suggests the enzyme holds the substrate in a position which strains and weakens the substrate's molecular bonds, making them easier to cleave and resulting in a general lowering of the energy of activation of the reaction. This postulate is extremely simplistic – the actual forces at work are much more numerous and complex.

When the substrate binds to the enzyme, it combines with only the active site. Information about the active site, such as its location and the nature and sequence of amino acids in it, provides an indication of the mechanism of binding and catalysis. The binding of the substrate to the enzyme's active site depends on many forces: hydrogen bonding, the interaction of hydrophobic (water-repelling) groups, and the electrostatic interaction between charged groups on the amino acids. Many active sites also contain metal ions which aid in binding the substrate or expediting the catalytic reaction by withdrawing or stabilizing electrons. For example, the enzyme carboxypeptidase, which hydrolyzes polypeptide bonds of proteins in food, contains a zinc atom in its active site. The electrophilic (electron-attracting) zinc atom coordinates electrons from the carbonyl of the peptide bond, weakening the bond for attack by a specific amino acid of the enzyme at the active site. Such a mechanism, however, is beyond the scope of elementary biology, and one would require a good course in biochemistry to understand fully.

Regulatory or allosteric enzymes have two binding sites: an active site and a regulatory site. Regulatory enzymes are a key controlling factor in metabolic pathways. If the end product of a pathway is in excess, it inhibits the action of the regulatory enzyme by binding to its regulatory site. The end product shuts off the catalytic activity of the active site by altering the arrangement of the enzyme's polypeptide chains, thus deforming and inactivating the enzyme (see the figure on the next page). This feedback mechanism is known as end-product inhibition and is important in preventing the accumulation of unwanted substances.

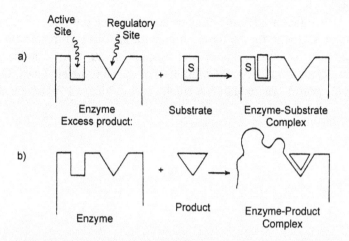

Active Site Regulatory Site

a)

Enzyme
Excess product:

Substrate

Enzyme-Substrate Complex

b)

Enzyme

Product

Enzyme-Product Complex

Schematic diagram showing binding at the active site (a) and regulatory site (b) of an enzyme. Note the change in enzyme conformation accompanying binding of product to the regulatory site.

There are several factors which affect enzyme activity, one of which is temperature. High temperatures of 50°C or above can inactivate or denature most enzymes. Upon denaturation, the structure of the enzyme is permanently altered resulting in an irreversible loss of activity. When most organisms are exposed to high temperatures, death occurs due to enzyme inactivation and the resulting loss of metabolic activity. Enzymes are usually not denatured by freezing, but their activity is decreased or disappears. This loss of activity is temporary and the activity reappears upon exposure to normal temperatures. Most enzymes have an optimal temperature range. At temperatures below 50°C, enzymatic reactions increase in proportion with each temperature increase.

Enzymes are also affected by and can be denatured by changes in pH. Although most enzymes have an optimum pH around neutrality (pH 7), some require an acidic medium and others require an alkaline medium. For example, both pepsin and trypsin work optimally at pH 8.5. The dependence of enzyme activity on pH is explained by the presence of ionizable groups on the protein molecules of the enzyme. pH controls, in part, the number of positive and negative charges on the enzyme molecule, which consequently affect activity.

A final factor affecting enzyme activity is the presence of enzyme poisons. Cytochrome oxidase, an enzyme involved in respiration, is inactivated by minute amounts of cyanide. Death from cyanide poisoning thus results from the inhibition of cytochrome enzymes. Other enzyme poisons include iodoacetic acid, fluoride, and lewisite.

CHAPTER 4

Nutrition in Plants and Animals

4.1 Importance of Digestion

After food has been ingested, it must be digested or broken down into smaller molecules so that these molecules can pass through plasma membranes and reach the cytoplasm inside the cells.

Carbohydrate molecules must be broken down to small molecules of simple sugars; protein molecules must be broken down into small molecules of amino acids; and lipid molecules must be broken down into molecules of fatty acids and glycerol.

Intracellular Digestion – In some heterotrophic organisms, digestion occurs after the solid material has actually been engulfed by a cell. This is common in the protozoans.

Extracellular Digestion – Organisms carry out digestion outside their cells, usually within the cavity of a digestive system; this is referred to as extracellular digestion. The simplest approach to this is employed by saprophytes who secure their food from nonliving but organic matter. This is the only type of digestion in many multicellular organisms, including humans.

In higher organisms, extracellular digestion takes place in two phases, mechanical and chemical.

4.1.1 Mechanical Phase of Digestion

Chewing and grinding break down food into smaller particles so that the total surface area of the food exposed to chemical action is increased.

4.1.2 Chemical Phase of Digestion

Chemical digestion, which is carried on by certain enzymes, reduces the size of the particles. By means of hydrolysis reactions, enzymes break the chemical bonds that hold the large molecules together.

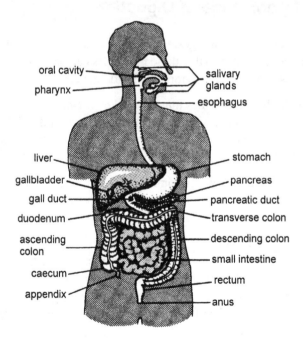

Figure 4.1 Human digestive system (the organs are slightly displaced, and the small intestine is greatly shortened)

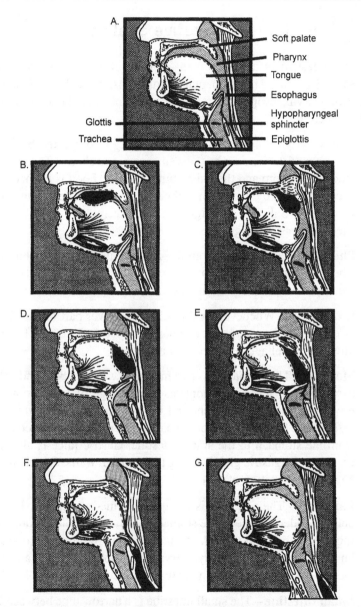

Figure 4.2 Movement of a bolus of food through the pharynx and upper esophagus during swallowing A) Mouth and pharynx at rest B) Early oral phase C) Late oral phase D) Early pharyngeal phase E) Middle pharyngeal phase F) Late pharyngeal phase G) Esophageal phase

4.2   The Human Digestive System

The digestive system of humans consists of the alimentary canal and several glands. This alimentary canal consists of the oral cavity (mouth), pharynx, esophagus, stomach, small intestine, large intestine, and the rectum.

A) **Oral Cavity (Mouth)** – The mouth cavity is supported by jaws and is bound on the sides by the teeth, gums, and cheeks. The tongue binds the bottom and the palate binds the top. Food is pushed between the teeth by the action of the tongue so it can be chewed and swallowed. Saliva is the digestive juice secreted that begins the chemical phase of digestion.

B) **Pharynx** – Food passes from the mouth cavity into the pharynx where the digestive and respiratory passages cross. Once food passes the upper part of the pharynx, swallowing becomes involuntary.

C) **Esophagus** – Whenever food reaches the lower part of the pharynx, it enters the esophagus and peristalsis pushes the food further down the esophagus into the stomach.

D) **Stomach** – The stomach has two muscular valves at both ends: the cardiac sphincter, which controls the passage of food from the esophagus into the stomach, and the pyloric sphincter, which is responsible for the control of the passage of partially digested food from the stomach to the small intestine. Gastric juice is also secreted by the gastric glands lining the stomach walls. Gastric juice begins the digestion of proteins.

E) **Pancreas** – The pancreas is the gland formed by the duodenum and the under surface of the stomach. It is responsible for producing pancreatic fluid which aids in digestion. Sodium bicarbonate, amylase, lipase, trypsin, chymotrypsin, carboxypeptidase, and nucleases are all found in the pancreatic fluid.

F) **Small Intestine** – The small intestine is a narrow tube between 20 and 25 feet long divided into three sections: the duodenum, jejunum, and ileum. The final digestion and absorption of disaccharides, peptides, fatty acids, and monoglycerides is the work of villi, which line the small intestine.

G) **Liver** – Even though the liver is not an organ of digestion, it does secrete bile which aids in digestion by neutralizing the acid chyme from the stomach and emulsifying fats. The liver is also responsible for the chemical destruction of excess amino acids, the storage of glycogen, and the breakdown of old red blood cells.

H) **Large Intestine** – The large intestine receives the liquid material that remains after digestion and absorption in the small intestine have been completed. However, the primary function of the large intestine is the reabsorption of water.

Table 4.1 Summary of the action of enzymes

Gland	Place of Action	Enzymes	Substrates	End Products
Salivary	Mouth	Ptyalin (amylase)	Starch	
Gastric	Stomach	Pepsin	Proteins (minerals)	(Dissolved minerals)
Liver	Small intestine	None		
Pancreas	Small intestine	Trypsin (protease)	Proteins	
		Amylopsin (amylase)	Starch	
		Steapsin (lipase)	Emulsified lipids	Fatty acids and glycerol
		Nucleases	Nucleic acids	Nucleotides
Intestinal	Small intestine	Peptidases	Polypeptides and dipeptides	Amino acids
		Maltase	Maltose	Glucose
		Sucrase	Sucrose	Glucose and fructose
		Lactase	Lactose	Glucose and galactose

Problem Solving Example:

Name the major organs of the human digestive tract and explain their functions.

The human digestive system begins at the oral cavity. The teeth break up food by mechanical means, increasing the surface area available to the action of digestive enzymes. There are four types of teeth. The chisel-shaped incisors are used for biting, while the pointed canines function in tearing, and the flattened, ridged premolars and molars are used for grinding and crushing food. In addition to tasting, the tongue manipulates food and forms it into a semi-spherical ball (bolus) with the aid of saliva.

The salivary glands consist of three pairs of glands. They produce watery and mucous saliva which coagulates food particles and lubricates the throat for the passage of the bolus. Saliva also contains amylases, which break down starches.

The tongue pushes the bolus into the pharynx which is the cavity where the esophagus and trachea (windpipe) meet. The larynx is raised against the epiglottis and the glottis is closed, preventing food from passing into the trachea. The act of swallowing initiates the movement of food down through a tube called the esophagus, which connects the mouth to the stomach. Once inside the esophagus, the food is moved by involuntary peristaltic waves toward the stomach.

The stomach is a thick muscular sac positioned on the left side of the body just beneath the ribs. The powerful stomach muscles break up the food, mix it with gastric juice and move it down the tract. Gastric juice is a mixture of hydrochloric acid and enzymes that further digest the food. Gastric juice and mucus are secreted by the small gastric glands in the lining of the stomach. The mucus helps protect the stomach from its own digestive enzymes and acid. The partially digested food, called chyme, is pushed through the pyloric sphincter into the small intestine.

The first part of the small intestine, called the duodenum, is held in a fixed position. The rest of the intestine is held loosely in place by a thin membrane called the mesentery, which is attached to the back of the body wall. In the duodenum, bile from the liver that has been stored in the gallbladder is mixed with pancreatic juice from the pancreas. The secretions of the pancreas and glandular cells of the intestinal tract contain enzymes that finish digesting the food. As digestion continues in the lower small intestine, muscular contractions mix the food and move it along. Small finger-like protrusions, called villi, line the small intestine facing the lumen. They greatly increase the intestinal surface area and absorb most of the nutrients.

The small intestine joins the large intestine (colon) at the cecum. The cecum is a blind sac that has the appendix protruding from one side. Neither the appendix nor the cecum are functional.

The large intestine has the function of removing water from the unabsorbed material. The last section of the colon stores feces until it is excreted through the anal sphincter.

 What is the basic mechanism of digestion, and what digestive processes take place in the mouth and stomach?

The process of digestion is the breakdown of large, ingested molecules into smaller, simple ones that can be absorbed and used by the body. The breakdown of these large molecules is called degradation. During degradation, some of the chemical bonds that hold the large molecules together are split. The digestive enzymes cleave molecular bonds by a process called hydrolysis, where a water molecule cleaves the bond.

Within living systems, chemical reactions require specific enzymes to act as catalysts. Enzymes are very specific, acting only on certain substrates. In addition, different enzymes work best under unlike conditions. Digestive enzymes work best outside of the cell, for their optimum pHs lie either on the acidic (e.g., gastric enzymes) or basic side (e.g., intestinal and pancreatic enzymes). The cell interior, however, requires an almost neutral (about 7.4) pH constantly. Digestive enzymes are secreted into the digestive tract by the cells that line it.

Digestion begins in the mouth. Most foods contain polysaccharides, such as starch, which are long chains of glucose molecules. Saliva (and the intestinal secretions) contains enzymes that degrade such molecules. Salivary amylase, an enzyme that is also called ptyalin, hydrolyzes starch into maltose. Glucose is eventually absorbed by the epithelial cells lining the small intestine.

The saliva has a pH of 6.5 – 7.5. This is the optimal range for salivary enzyme activity. Food spends a relatively short amount of time in the mouth, and eventually enters the stomach. The stomach is very acidic, with a pH of 1.5-2.5. The low pH is required for the activity of the stomach enzyme pepsin. Pepsin is a proteolytic enzyme, which degrades proteins. Pepsin starts the protein digestion in the stomach by splitting the long proteins into shorter fragments, or peptides, that are further digested in the intestine. There are 20 different kinds of amino acids that can make up a protein and some proteins are thousands of amino acids long. The body needs the amino acids it obtains from digestion to synthesize its own proteins.

4.3 Ingestion and Digestion in Other Organisms

A) **Hydra –** The hydra possesses tentacles that have stinging cells (nematocysts) which shoot out a poison to paralyze the prey. If successful in capturing an animal, the tentacles push it into the hydra's mouth. From there, the food enters the gastric cavity. The hydra uses both intracellular and extracellular digestion.

B) **Earthworm –** As the earthworm moves through soil, the suction action of the pharynx draws material into the mouth cavity. Then from the mouth, food goes into the pharynx, the esophagus, and then the crop which is a temporary storage area. This food then passes into a muscular gizzard where it is ground and churned. The food mass finally passes into the intestine; any undigested material is eliminated through the anus.

C) **Grasshopper –** The grasshopper is capable of consuming large amounts of plant leaves. This plant material must first pass through

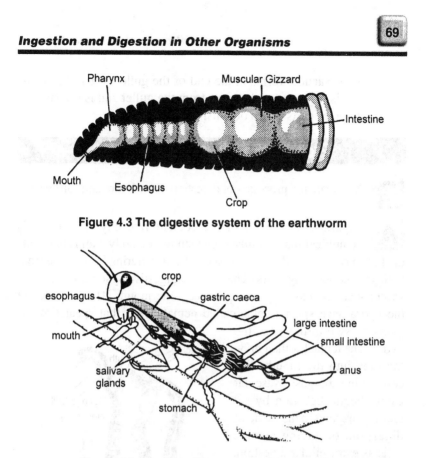

Figure 4.3 The digestive system of the earthworm

Figure 4.4 The digestive system of the grasshopper

the esophagus into the crop, a temporary storage organ. It then travels to the muscular gizzard where food is ground. Digestion takes place in the stomach. Enzymes secreted by six gastric glands are responsible for digestion. Absorption takes place mainly in the stomach. Undigested material passes into the intestine, collects in the rectum and is eliminated through the anus.

D) Protozoa

   1) **Amoeba** – The amoeba ingests food by means of temporary pseudopods. These pseudopods engulf a food particle and form a food vacuole within the cytoplasm.

   2) **Paramecium** – With the aid of cilia, the paramecium sweeps food particles into the oral groove and then into the gullet. A

food vacuole forms at the end of the gullet; when this vacuole is filled, it breaks away from the gullet and is transported to other parts of the organism.

Q Compare the processes of digestion in the hydra and earthworm.

A Hydra is a member of the coelenterate phylum. Hydra is a multicellular, sac-like organism whose body wall is two cells thick. Many of the cells are semi-specialized, differing in function from their immediate neighbors. The hydra's digestive system, the gastrovascular cavity, has only one opening – the mouth. The tentacles of the hydra have special cells, called nematocysts, that spear tiny organisms swimming near the hydra. The tentacles then draw the prey into the mouth. The cells lining the gastrovascular cavity begin digestion by secreting enzymes. This mode of digestion, occurring outside cells, is extracellular digestion, which is followed by intracellular digestion. When small food fragments come into contact with the cells lining the cavities, these fragments are phagocytized and digestion is completed inside the cell. The indigestible material is expelled through the mouth. Thus, the hydra utilizes extracellular digestion to break down relatively large organisms into small pieces which are then digested intracellularly by individual cells.

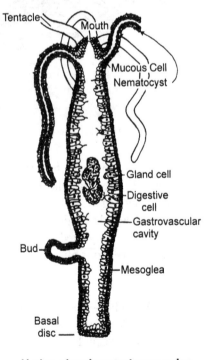

Hydra, showing gastrovascular cavity with food material in it

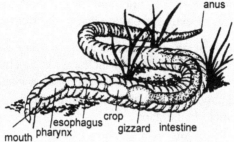

Digestive system of an earthworm

An earthworm has a complete digestive tract. It has two openings, the mouth and anus, with wastes expelled through the anus and food ingested through the mouth. Various regions of the tract are specialized: the pharynx sucks in food, the esophagus transports food to the crop, the crop stores the food, the gizzard breaks the food up, and the intestine digests and absorbs the food. Most digestion is extracellular, utilizing the enzymes secreted by cells lining the tract. The digestive tract is not widely branched, but is rather like a simple gastrovascular cavity. However, large numbers of folds are found in the tract lining, which greatly increase the absorptive surface area. Since the earthworm is a land animal, it must also conserve its body fluids by reabsorbing water from the indigestible wastes. This occurs in the posterior region of the intestine, analogous to the human large intestine.

4.4 Transport of Food in Vascular Plants

A) **Leaf** – The leaf consists of an upper epidermis, a mesophyll layer, and a lower epidermis. Its primary function is to change inorganic substances to organic substances by the process of photosynthesis. A leaf also functions in the exchange of gases between the plant and the atmosphere.

The veins in the leaf, known as xylem and phloem tubes, transport fluid materials in the leaf.

1) **Xylem** – carries water in an upward direction and dissolved minerals from the stem and roots to the leaf cells.

2) **Phloem** – transports food materials, in both an upward and downward direction, from the leaf to the stem and roots.

B) **Stem** – The stem consists of an epidermis, sclerenchyma, parenchyma, and conducting tissue which are the xylem and phloem. One of the primary functions of the stem is to transport raw materials from the roots to the leaves and manufactured products to the roots and other plant organs.

C) **Root** – The water that is needed by the plants enters by way of the roots. Water and dissolved minerals diffuse into the root hairs and pass through the cortex cells to the cells of the xylem.

Problem Solving Example:

 What are the functions of the xylem and the phloem?

 In vascular plants, the vital function of the transportation of food, water, and minerals is performed by the vascular system, composed of the xylem and the phloem. The xylem is chiefly concerned with the conduction of water and mineral salts from the roots to the above-ground portion of the plant, where they are used for photosynthesis and other metabolic purposes. The xylem also serves as a means of support in larger vascular plants. Both are elongated cells with thickened cellulose walls heavily impregnated with lignin; both contain no living protoplasm, and hence are dead cells.

Carbohydrates are manufactured primarily in the leaves of a plant. They are transported to the other parts of the plant by the phloem, which runs parallel to the xylem throughout the plant body. The phloem consists of elongated, tube-like cells, with specialized pores at each end. They are living and have a protoplasm, but no nuclei. Besides conduction, the phloem serves as a supporting tissue, due to the presence of strong fibers in the walls of its cells.

Quiz: Cellular Metabolism – Nutrition

1. Which of the following is concerned mainly with cellular respiration?

    (A) Cell membrane

    (B) Golgi apparatus

    (C) Ribosomes

    (D) Mitochondria

2. In comparing photosynthesis and cellular respiration, which one of the following statements would not be true?

    (A) ATP is formed during both processes.

    (B) CO_2 is produced during both processes.

    (C) O_2 is released during photosynthesis only.

    (D) Several enzymes are needed for each process to occur.

3. The reduced form of the coenzyme in the dehydrogenase enzymes is

    (A) NADH.

    (B) FAD.

    (C) ADH.

    (D) NAD^+.

4. Which carbohydrate can humans not digest?

    (A) Starch

    (B) Fructose

    (C) Cellulose

    (D) Maltose

5. How many ATPs are derived from the oxidation of 1 molecule of pyruvate via the Krebs cycle and the electron transport system?

    (A) 12

    (B) 14

    (C) 15

    (D) 20

6. Which of the following contributes to the "powerhouse" properties of the mitochondria?

    (A) The presence on the mitochondria's inner membrane of the enzymes necessary for the functioning of the electron transport system

    (B) The presence of cristae, which increase the surface area of the mitochondria's inner membrane

    (C) The ease with which pyruvate can travel through the outer and inner membrane into the mitochondria's inner matrix

    (D) All of the above

7. Which of the following statements is *incorrect* concerning digestion?

   (A) Although very little digestion actually occurs in the stomach, some cells lining the stomach do secrete hydrochloric acid, which aids in the breakdown of proteins by activating a proteolytic enzyme.

   (B) Most digestion takes place in the upper $\frac{1}{3}$ of the small intestine, an area known as the duodenum.

   (C) Bile, in addition to pancreatic enzymes, is secreted into the cecum to put the "finishing touches" on digestion.

   (D) If food is not digested properly when it reaches the large intestine, it can cause severe problems, as the bacteria there can ferment the leftover carbohydrates, multiply rapidly, and cause infection.

8. Which of the following is *not* a proteolytic enzyme?

   (A) Trypsin

   (B) Lipase

   (C) Carboxypeptidase

   (D) Pepsin

9. Which of the following is *not* a function of hydrochloric acid secreted into the stomach?

   (A) As a bacteriocidal agent

   (B) As a "turn-on" factor for some proteolytic enzymes

   (C) As an aid in the formation of chylomicrons for fat digestion

   (D) As a denaturing agent

10. Which of the following is *incorrect* concerning the pancreas?

   (A) Among its secretion products are the enzymes trypsin and chymotrypsin.

   (B) The pancreas plays only a minor role in blood sugar-level regulation.

   (C) The digestive secretions of the pancreas are controlled in part by the hormone secretin.

   (D) The pancreas can be considered an exocrine and an endocrine gland.

ANSWER KEY

1.	(D)	6.	(D)
2.	(B)	7.	(C)
3.	(A)	8.	(B)
4.	(C)	9.	(C)
5.	(C)	10.	(B)

CHAPTER 5

Gas Exchange in Plants and Animals

5.1 Respiration in Humans

The respiratory system in humans begins as a passageway in the nose. Inhaled air then passes through the pharynx, trachea, bronchi, and lungs.

A) **Nose** – The nose is better adapted to inhale air than the mouth. The nostrils, the two openings in the nose, lead into the nasal passages which are lined by the mucous membrane. Just beneath the mucous membrane are capillaries which warm the air before it reaches the lungs.

B) **Pharynx** – Air passes via the nasal cavities to the pharynx where the paths of the digestive and respiratory systems cross.

C) **Trachea** – The upper part of the trachea, or windpipe, is known as the larynx. The glottis is the opening in the larynx; the epiglottis, which is located above the glottis, prevents food from entering the glottis and obstructing the passage of air.

D) **Bronchi** – The trachea divides into two branches called the bronchi. Each bronchus leads into a lung.

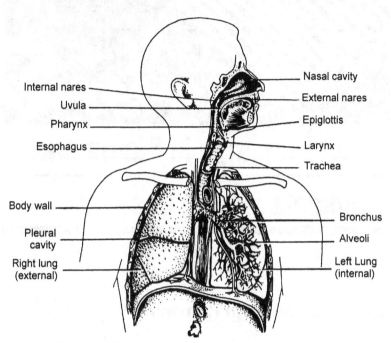

Figure 5.1 Diagram of the human respiratory system

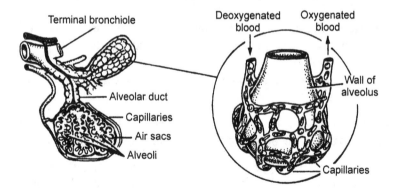

Figure 5.2 Diagram of a small portion of the lung, highly magnified, showing the air sacs at the end of the alveolar ducts, the alveoli in the walls of the air sacs, and the proximity of the alveoli and the pulmonary capillaries containing red blood cells

E) **Lungs** – In the lungs, the bronchi branch into smaller tubules known as the bronchioles. The finer divisions of the bronchioles eventually enter the alveoli. The cells of the alveoli are the true respiratory surface of the lung. It is here that gas exchange takes place.

5.1.1 Stages of Respiration

A) **External Respiration** – This stage of respiration involves the exchange of gases between the air sacs and the blood stream, and breathing movements (inhalation and exhalation).

B) **Internal Respiration** – This stage of respiration involves the exchange of gases between the blood and the body cells.

C) **Cellular Respiration** – This stage of respiration takes place within cells where both aerobic and anaerobic oxidation reactions take place.

Problem Solving Examples:

 Differentiate clearly between "breathing" and "respiration."

 Respiration has two distinct meanings. It refers to the oxidative degradation of nutrients such as glucose through metabolic reactions within the cell, resulting in the production of carbon dioxide, water, and energy. Respiration also refers to the exchange of gases between the cells of an organism and the external environment. Many different methods for exchange are utilized by different organisms. In man, respiration can be categorized by three phases: ventilation (breathing), external respiration, and internal respiration.

Breathing may be defined as the mechanical process of taking air into the lungs (inspiration) and expelling it (expiration). It does not include the exchange of gases between the bloodstream and the alveoli. Breathing must occur in order for respiration to occur; that is, air must be brought to the alveolar cells before exchange can be effective. One distinction that can be made between respiration and breathing is that the former ultimately results in energy production in the cells. Breathing, on the other hand, is solely an energy consuming process because of the muscular activity required to move the diaphragm.

 List the parts of the human respiratory system. How is each adapted for its particular function?

The respiratory system in man and other air-breathing vertebrates includes the lungs and the tubes by which air reaches them. Normally, air enters the human respiratory system by way of nostrils, but it may also enter by way of the mouth. The nostrils, which contain small hairs to filter incoming air, lead into the nasal cavities, which are separated from the mouth below by the palate. The nasal cavities contain the sense organs of smell, and are lined with mucus-secreting epithelium which moistens the incoming air. Air passes from the nasal cavities via the internal nares into the pharynx, then through the glottis and into the larynx. Stretched across the larynx are the vocal cords. The opening to the larynx, called the glottis, is always open except in swallowing, when a flap-like structure (the epiglottis) covers it. Leading from the larynx to the chest region is a long cylindrical tube called the trachea, or windpipe. In the middle of the chest, the trachea bifurcates into bronchi which lead to the lungs. In the lungs, each bronchus branches, forming smaller and smaller tubes called bronchioles. The smaller bronchioles terminate in clusters of cup-shaped cavities, the air sacs. In the walls of the smaller bronchioles and the air sacs are the alveoli, which are moist structures supplied with a rich network of capillaries. Internal gas exchange respiration takes place when molecules of oxygen and carbon dioxide diffuse readily through the thin, moist walls of the alveoli.

The chest cavity is closed and has no communication with the outside. It is bounded by the chest wall, which contains the ribs on its top, sides and back, and the sternum anteriorly. The bottom of the chest wall is covered by a strong, dome-shaped sheet of skeletal muscle, the diaphragm. The diaphragm separates the chest region (thorax) from the abdominal region, and plays a crucial role in breathing by contracting and relaxing, changing the intrathoracic pressure.

5.2 Respiration in Other Organisms

A) Protozoa

1) **Amoeba** – Simple diffusion of gases between the cell and water is sufficient to take care of the respiratory needs of the amoeba.

2) **Paramecium** – The paramecium takes in dissolved oxygen and releases dissolved carbon dioxide directly through the plasma membrane.

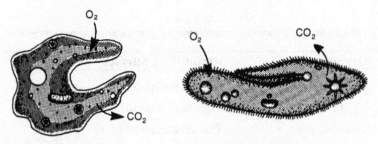

Figure 5.3 Respiration in the amoeba

Figure 5.4 Respiration in the paramecium

B) **Hydra** – Dissolved oxygen and carbon dioxide diffuse in and out of two cell layers through the plasma membrane.

C) **Grasshopper** – The grasshopper carries on respiration by means of spiracles and hollow tracheae. Blood plays no role in transporting oxygen and carbon dioxide. Muscles of the abdomen pump air into and out of the spiracles and the tracheae.

D) **Earthworm** – The skin of the earthworm is its respiratory surface. Oxygen from the air diffuses into the capillaries of the skin and joins with hemoglobin dissolved in the blood plasma. This oxyhemoglobin is released to the tissue cells. Carbon dioxide from the tissue cells diffuses into the blood. When the blood reaches the capillaries in the skin again, the carbon dioxide diffuses through the skin into the air.

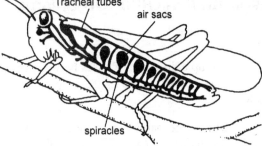

Figure 5.5 Respiration in the grasshopper

Table 5.1
Comparison of various respiratory surfaces among organisms

Organism	Respiratory Surface
Protozoan	Plasma membrane
Hydra	Plasma membrane of each cell
Grasshopper	Tracheae network
Earthworm	Moist skin
Human	Air sacs in lungs

Problem Solving Examples:

 How does respiration occur in a grasshopper?

 The unique respiratory system of insects consists of a network of tubes called tracheae. These tracheae open to the outside by means of holes in the insect's body. These holes are called spiracles and are conspicuous on the sides of the thorax and abdomen. Each spiracle is guarded by a valve which can be opened or closed to regulate air flow. The tracheal tubes extend to all the internal organs. Oxygen and carbon dioxide diffuse through these into the adjacent cells. The body wall of the insect pulsates, drawing air into the trachea when the body expands, and forcing air out when the body contracts. Grasshoppers draw air into the body through the first four pairs of spiracles when the abdomen expands and expel it through the last six pairs when the body contracts. Therefore, in contrast to a fish or a crab which respires through gills, the tracheal system conducts air deep within the insect's body. The air is brought near enough to each cell so that gases can diffuse across the wall of the tracheal tube. For this reason, insects need not maintain a rapid blood flow, as vertebrates must, to supply their cells with oxygen.

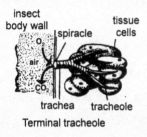

Terminal tracheole

 Compare the methods of obtaining oxygen in the earthworm and humans.

The earthworm belongs to the group of invertebrates called the Annelids, or roundworms. The earthworm has no specialized respiratory system. Respiration takes place by diffusion of gases through the body integument or skin. The skin is kept moist by the secretions of mucous glands directly underneath the skin. The thin cuticle is quite permeable to both oxygen and carbon dioxide. The red blood cells contain dissolved hemoglobin, which aids in the transportation of oxygen to various parts of the body.

Gas exchange in humans takes place in the lungs. Lungs serve the same purpose in terrestrial animals as gills do in aquatic animals. The lungs are located internally in order to prevent desiccation and possible damage.

It should be noted that as an animal becomes larger and more complex, and more metabolic energy is needed, the surface area for gas exchange likewise increases. This facilitates the uptake of oxygen in addition to the release of carbon dioxide. The extensive branching of the bronchioles into alveoli in the lungs are adaptations toward increased surface area for gas exchange.

5.3   Gas Exchange in Plants

A) Gas Exchange in Roots and Stems

Plants are able to exist without specialized organs for gas exchange because:

1) there is little transport of gases from one part of the plant to another.

2) plants respirate at a much lower rate than animals.

3) the distance gas must diffuse is not large.

4) most cells have at least a portion of their outer surface in contact with air.

B) Gas Exchange in the Leaf

The exchange of gases in the leaf for photosynthesis occurs through pores in the surface of the leaf known as stomata. Usually, the stomata open during the presence of light and close in its absence. The most direct cause of this is the change of turgor in the guard cells.

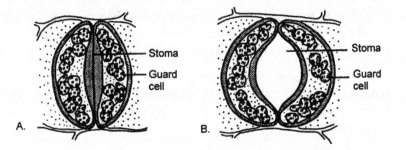

Figure 5.6   A. Stomata closed
B. Stomata open when turgor builds in guard cells

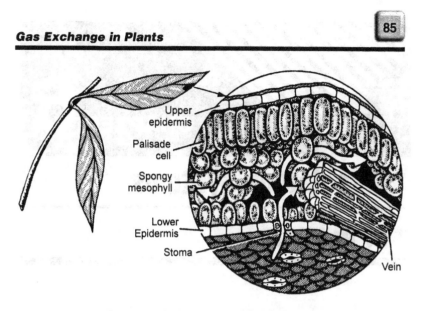

Figure 5.7 A diagram representing loosely arranged cells in a leaf which allow for rapid diffusion of gas

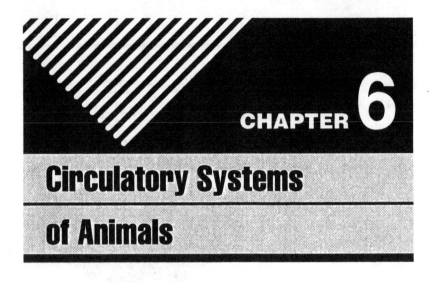

CHAPTER 6

Circulatory Systems

of Animals

6.1 The Human Circulatory System

Humans have a closed circulatory system in which the blood moves entirely within the blood vessels. This circulatory system consists of the heart, blood, veins, arteries, capillaries, lymph, and lymph vessels.

A) Heart

The heart is a pump covered by a protective membrane known as the pericardium and divided into four chambers. These chambers are the left and right atria and the left and right ventricles.

1) **Atria** – The atria are the upper chambers of the heart that receive blood from the superior and inferior vena cava. This blood is then pumped to the lower chambers or ventricles.

2) **Ventricles** – The ventricles have thick walls as compared to the thin walls of the atria. They must pump blood out of the heart to the lungs and other distant parts of the body.

The heart also contains many important valves. The tricuspid valve is located between the right atrium and the right ventricle. It prevents the backflow of blood into the atrium after the contraction of the right ventricle. The bicuspid valve, or mitral valve, is situated between the left atrium and the left ventricle. It prevents the backflow of blood into the left atrium after the left ventricle contracts.

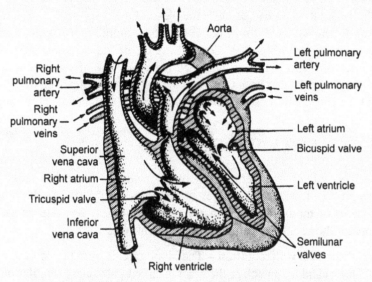

Figure 6.1 The functional parts of the heart and direction of blood flow through its chambers

Problem Solving Example:

 Explain why a four-chambered heart is more efficient than a three-chambered heart.

 A four-chambered heart is characteristic of endothermic, or "warm-blooded" animals such as birds and mammals. Since these animals maintain a relatively high constant body temperature, they must have a fairly high metabolic rate. To accomplish this, much oxygen must be continually provided to the body's tissues.

A four-chambered heart helps to maximize this oxygen transport by keeping the oxygenated blood completely separate from the deoxygenated blood. The right side of the heart, which carries the deoxygenated blood, is separated by a muscular wall, called the septum, from the left side of the heart, which carries the oxygenated blood. In amphibians, the atria are divided into two separate chambers, but a single ventricle exists. This three-chambered heart permits oxygen-rich blood returning from the pulmonary circulation to mix with oxygen-poor blood returning from the systemic circulation. It is less efficient than a four-chambered heart because the blood flowing to the tissues is not as oxygen-rich as it could be. Fortunately, amphibians, being ectothermic, or "cold-blooded," do not have to maintain a constant body temperature and hence do not need the efficiency of warm-blooded hearts. Reptiles also have three-chambered hearts, but partial division of the ventricle in these animals has decreased the amount of mixing.

6.2   Blood Circuits

The pulmonary artery carries blood from the heart to the lungs to oxygenate blood. The branches of the aorta carry blood to the other systems of the body. Semilunar valves, at the base of these arteries, prevent the backflow of blood into the ventricles.

A) **Pulmonary Circulation** – The blood circuit from the heart to the lungs and then back to the heart is called pulmonary circulation.

B) **Systemic Circulation** – In this blood circuit, blood circulates from the left ventricle into the aorta and then to all body systems, except the lungs, and back to the heart. There are three branches of systemic circulation. These are the coronary circulation, the hepatic portal system, and renal circulation.

C) **Veins** – All the veins in the body carry blood to the heart. Veins possess small amounts of muscle tissue in their walls and, therefore, the walls are thinner. Valves, which prevent the backflow of blood, are found along the length of the vein. Every vein in the body, except the pulmonary vein, carries deoxygenated blood.

D) **Arteries** – Every artery in the body, except the pulmonary artery, carries oxygen-rich blood. Every artery in the body carries blood away from the heart. Unlike the thin walls of veins, arteries possess relatively thick walls.

E) **Capillaries** – The capillaries, microscopic in size, are the smallest and most abundant blood vessels of the body. They connect arterioles and venules to complete the circuit of blood. Capillaries function in the exchange of material between the cells and the blood. The walls of the capillary consist of a single layer of flattened epithelial cells.

F) **Lymph** – Lymph is the intercellular fluid present around every internal body cell. It is composed of water and the dissolved substances that pass out of the capillaries and the tissue cells. Lymph is the medium through which materials are exchanged between the tissue cells and blood vessels.

G) **Lymph Vessels** – The lymphatic vessels function to return fluid from the tissue spaces to the circulation. Lymphatic vessels join larger collecting lymphatics such as the thoracic duct. The thoracic duct carries lymph upward across the chest and empties into a large

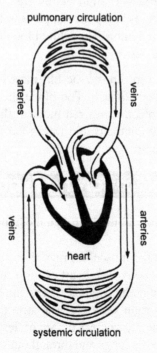

Figure 6.2 Schematic representation showing pulmonary and systemic circulation

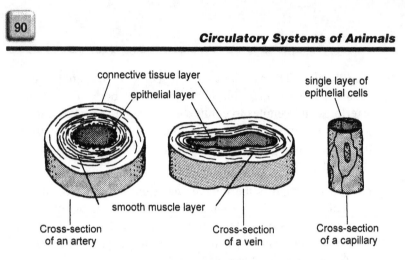

connective tissue layer

epithelial layer

single layer of
epithelial cells

smooth muscle layer

Cross-section
of an artery

Cross-section
of a vein

Cross-section
of a capillary

Figure 6.3 Comparison of blood vessels

vein near the base of the neck. Lymph travels in only one direction, from the body organs to the heart.

1) **Lymph Nodes** – Lymph nodes are the thousands of pea-shaped structures lining the lymph vessels. They filter out bacteria from the lymph. The white blood cells produced by the lymph nodes destroy bacteria.

2) **Spleen** – The spleen is a sac-like mass of lymphatic tissue located near the stomach. The blood that passes into it is filtered. The spleen stores red blood cells so that when bleeding occurs, it contracts and forces the stored red blood cells into circulation.

Problem Solving Example:

 Trace the path of blood through the human heart.

 The heart is the muscular organ that causes the blood to circulate in the body. The heart of birds and mammals is a pulsatile four-chambered pump composed of an upper left and right atrium and a lower left and right ventricle. The atria function mainly as entry ways to the ventricles; whereas the ventricles supply the main force that propels blood to the lungs and throughout the body.

Depending from where the blood is flowing, it would enter the heart via one of two veins: the superior vena cava carries blood from the head, neck and arms; the inferior vena cava carries blood from the rest of the body. The blood from these two veins enters the right atrium. When this chamber is filled with blood, the chamber contracts and forces the blood through the tricuspid valve and into the right ventricle. Since this blood has returned from its circulation in the body's tissues, it is deoxygenated and contains much carbon dioxide. It therefore must be transported to the lungs where gas exchange can take place. The right ventricle contracts, forcing the blood through the pulmonary semilunar valve into the pulmonary artery. This artery is unlike most arteries in that it carries deoxygenated blood. The artery splits into two, with one branch leading to each lung. The pulmonary arteries further divide into many arterioles, which divide even further and connect with dense capillary networks surrounding the alveoli in the lungs. The alveoli are small sac-like cavities where gas exchange occurs. Carbon dioxide diffuses into the alveoli, where it is expelled, while oxygen is picked up by the hemoglobin of the erythrocytes. The capillaries join to form small venules which further combine to form the four pulmonary veins leading back to the heart. The pulmonary veins are unlike most veins in that they carry oxygenated blood. These veins empty into the left atrium, which contracts to force the blood through the bicuspid (or mitral) valve into the left ventricle. When the left ventricle, filled with blood, contracts, the blood is forced through the aortic semilunar valve into the aorta, the largest artery in the body (about 25 millimeters in diameter).

The aorta forms an arch and runs posteriorly and inferiorly along the body. Before it completes the arch, the aorta branches into the coronary artery, which carries blood to the muscular walls of the heart itself; the carotid arteries, which carry blood to the head and brain; and the subclavian arteries, which carry blood to the arms. As the aorta runs posteriorly, it branches into arteries which lead to various organs such as the liver, kidney, intestines, spleen, and also the legs.

The arteries divide into arterioles which further divide and become capillaries. It is here that the oxygen and nutrients diffuse into the tissues, and carbon dioxide and nitrogenous wastes are picked up. The capillaries fuse to form venules which further fuse to become either the superior or inferior vena cava. The entire cycle starts once again.

The part of the circulatory system in which deoxygenated blood is pumped to the lungs and oxygenated blood returned to the heart is called the pulmonary circulation. The part in which oxygenated blood is pumped to all parts of the body by the arteries and deoxygenated blood is returned to the heart by the veins is called the systemic circulation.

6.3 The Components of Blood

A) **Plasma** – Plasma is the liquid part of the blood which constitutes 55% of the total blood volume. Blood plasma is essential for homeostasis.

B) **Blood Cells** – These constitute 45% of the blood.

1) **Red Blood Cells (Erythrocytes)** – The erythrocytes are the most numerous of the three blood cell types. Formed in the marrow of the bones, the red blood cells first possess a nucleus and not very much hemoglobin. As they mature, erythrocytes lose their cell nuclei and hemoglobin constitutes 90% of their dry weight. The chief function of erythrocytes is to carry oxygen to all parts of the body and to remove some carbon dioxide.

2) **White Blood Cells (Leukocytes)** – A white blood cell has a nucleus but lacks hemoglobin. They are larger than red blood cells and generally function to protect the body against disease. There are several types of white blood cells. These are the neutrophils, monocytes, lymphocytes, eosinophils, and basophils. Some are formed in certain bones, others, in the lymph nodes.

3) **Platelets** – These are cell fragments which are produced by large cells in the bone marrow. Platelets are much smaller than red or white blood cells. They play an important role in blood clotting.

Problem Solving Example:

Explain why blood is so important to many animals. Discuss the major functions of blood.

All cells, in order to survive, must obtain the necessary raw materials for metabolism, and have a means for the removal of waste products. In small plants and animals living in an aquatic environment, these needs are provided for by simple diffusion. The cells of such organisms are very near the external watery medium, and so nutrients and wastes do not have a large distance to travel. However, as the size of the organism increases, more and more cells become further removed from the media bathing the peripheral cells. Diffusion cannot provide sufficient means for transport. In the absence of a specialized transport system, the limit on the size of an aerobic organism would be about a millimeter, since the diffusion of oxygen and nutrients over great distances would be too slow to meet the metabolic needs of all the cells of the organism. In addition, without internal transport, organisms are restricted to watery environments, since the movement to land requires an efficient system for material exchange in non-aqueous surroundings. Therefore, larger animals have developed a system of internal transport, the circulatory system. This system, consisting of an extensive network of various vessels, provides each cell with an opportunity to exchange materials by diffusion.

Blood is vital to the circulatory system, transporting nutrients and oxygen to all the cells and removing carbon dioxide and other waste. Blood also serves other important functions. It transports hormones and the secretions of the endocrine glands, which affect organs sensitive to them. Blood also acts to regulate the acidity and alkalinity of the cells via control of their salt and water content. In addition, the blood acts to regulate the body temperature by cooling certain organs and tissues when an excess of heat is produced (such as in exercising muscle) and warming tissues where heat loss is great (such as in the skin).

Some components of the blood act as a defense against bacteria, viruses, and other pathogenic (disease-causing) organisms. The blood also has a self-preservation system called a clotting mechanism so that loss of blood due to vessel rupture is reduced.

6.4   The Functions of Blood

Blood:

A) Transports materials to and from all the tissues of the body.

B) Defends the body against infectious diseases.

Problem Solving Examples:

 What makes the blood flow throughout the body?

 Blood flow is the quantity of blood that passes a given point in the circulation in a given period of time. It is usually expressed in milliliters or liters per minute. Blood flow through a blood vessel is determined by two factors: (1) the pressure difference that drives the blood through the vessel; and (2) the vascular resistance or impedance to blood flow.

As the pressure difference increases, flow increases; as the resistance increases, flow decreases. The pressure difference between the two ends of the vessel causes blood to flow from the high pressure end to the low pressure end. This pressure difference, or gradient, is the essential factor which causes blood to flow. Although the heart acts to pump the blood, it is the pressure difference that is critical for blood flow. The overall blood flow in the circulation of an adult is about 5 liters per minute. This is called the cardiac output because it is the amount pumped by each ventricle of the heart per unit of time. The normal cardiac output is thus about 5 liters per minute.

 How are nutrients and waste materials transported from the circulatory system to the body's tissues?

 It is in the capillaries that the most important function of circulation occurs; that is, the exchange of nutrients and waste materials between the blood and tissues. There are billions of capillaries, providing a total surface area of over 100 square meters for the exchange of material. Most functional cells in the body are never more than about 25 microns away from a capillary.

Capillaries are well-suited for the exchange process. They are both numerous and branch extensively throughout the body. Unlike the arteries and veins, the capillaries have a very thin wall; it is one endothelial cell thick. This thin wall permits rapid diffusion of substances through the capillary membrane. The extensive branching increases the total cross-sectional area of the capillary system. This serves to slow down the flow of blood in the capillaries, allowing more time for the exchange process to occur. The very small diameter of each of the capillaries provides friction, which increases the resistance to flow. This causes a significant drop in blood pressure in the capillaries, which is important in the filtration of fluid between the capillaries and the interstitial fluid (lying in the space between cells).

The most important means by which nutrients and wastes are transferred between the plasma and interstitial fluid is by diffusion. Material must first diffuse into the interstitial fluid before it can enter the cells of the body's tissues. As the blood flows slowly through the capillary, large amounts of water molecules and dissolved substances diffuse across the capillary wall, proportional to the concentration difference between the two sides of the capillary membrane.

Actual diffusion across the capillary wall occurs in three ways. Water-soluble substances (sodium ions, glucose) diffuse between the plasma and interstitial fluid only through the pores in the capillary membrane. The diameter of the pores allows small molecules such as water, urea, and glucose to pass through, but not larger molecules such as plasma proteins. Lipid-soluble substances diffuse directly through the cell membranes and do not go through the pores (which are filled with water). Oxygen and carbon dioxide permeate all areas of the capillary membranes. Another method by which substances can be transported through the membrane is pinocytosis. In pinocytosis, the cell membrane invaginates and sequesters the surrounding substances. Small vesicles are produced which migrate from one side of the endothelial cell to the other, where the contents are released. Pinocytosis accounts for the transport of larger substances, such as plasma proteins and glycoproteins.

6.5   The Clotting of Blood

The sequence of events in blood clotting is initiated with the breakdown of platelets. With this, thromboplastin is liberated. In the presence of excess thromboplastin and calcium ions, the clotting enzyme prothrombin is changed to thrombin. This thrombin then acts on fibrinogen, a dissolved protein, and converts it into fibrin. At the wound site, the fibrin serves to trap the blood cells. A clot is soon formed which stops loss of blood from the severed vessel.

1)   Platelets + Damaged Cells $\xrightarrow{\text{Clotting Factors in Blood}}$ Thromboplastin

2)   Prothrombin $\xrightarrow[\text{Ca}^{++}]{\text{Thromboplastin}}$ Thrombin

3)   Fibrinogen $\xrightarrow{\text{Thrombin}}$ Fibrin

4)   Fibrin + Cells + Dried Serum $\longrightarrow$ Blood Clot

Figure 6.4 The sequence in blood clotting

Problem Solving Example:

 In an earlier question, we learned about the various factors which are involved in the clotting mechanism. Show how these factors are affected in conditions that cause excessive bleeding in humans and in conditions that cause intravascular clotting.

 Excessive bleeding can result from a deficiency of any one of the blood clotting factors. An insufficiency of prothrombin, one of the intermediates in the clotting mechanism, can cause a patient to develop a severe tendency to bleed. Both hepatitis and cirrhosis (diseases of the liver) can depress the formation of prothrombin in the liver. Vitamin K deficiency also depresses the levels of prothrombin. Vitamin K deficiency does not result from the absence of the vitamin from the diet, since it is continually synthesized in the gastrointestinal tract by bacteria. The deficiency results from poor absorption of fats (vitamin K is fat soluble) from the tract due to a lack of bile, which is secreted by the liver.

Hemophilia, or "bleeder's disease," is a term loosely applied to several different hereditary deficiencies in coagulation, resulting in bleeding tendencies. Most cases of hemophilia result from a deficiency in one of the protein factors called Factor VIII or the antihemophilic factor necessary for production of prothrombinase by the platelets.

Thrombocytopenia, having a very low quantity of platelets in the blood, also causes excessive bleeding. With a deficiency of platelets, not enough prothrombinase can be synthesized. One major type of thrombocytopenia results from the development of immunity to one's own platelets. The antibodies attack and destroy the platelets in the person's own blood. Thrombocytopenia also results from pernicious anemia and certain drug therapies.

These abnormalities cause excessive bleeding, but there are also pathological conditions caused by clotting when it should not normally occur. An abnormal clot that develops in a blood vessel is called a thrombus. If the thrombus breaks away from its attachment and flows freely in the bloodstream, it is termed an embolus. Should an embolus block an important blood vessel (to the heart, lungs, or brain), death could occur. Any roughened inner surface of a blood vessel, which may result from arteriosclerosis or physical injury, can initiate the clotting mechanism by releasing thromboplastin from the platelets. Blood which flows too slowly may also cause clotting. Since small amounts of thrombin are always being produced, a hampered flow can increase the concentration of thrombin in a specific area, so that a thrombus results. The immobility of bed patients presents this problem.

6.6 Transport Mechanisms in Other Organisms

A) **Protozoans** – Most protozoans are continually bathed by food and oxygen because they live in water or another type of fluid. With the process of cyclosis or diffusion, digested materials and oxygen are distributed within the cell, and water and carbon dioxide are removed. Proteins are transported by the endoplasmic reticulum.

B) **Hydra** – Like the protozoan, materials in the hydra are distributed to the necessary organelles by diffusion, cyclosis, and the endoplasmic reticulum.

C) **Earthworm** – The circulatory system of the earthworm is known as a "closed" system because the blood is confined to the blood vessels at all times. A pump that forces blood to the capillaries consists of five pairs of aortic loops. Contraction of these loops forces blood into the ventral blood vessel. This ventral blood vessel transports blood toward the rear of the worm. The dorsal blood vessel forces blood back to the aortic loops at the anterior end of the worm.

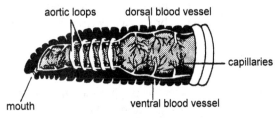

Figure 6.5 "Closed" circulatory system of the earthworm.

D) **Grasshopper** – The grasshopper possesses an "open" circulatory system where the blood is confined to vessels during only a small portion of its circuit through the body. The blood is pumped by the contractions of a tubular heart and a short aorta with an open end. Blood from the heart flows into the aorta and then into sinuses. The blood then returns to the heart.

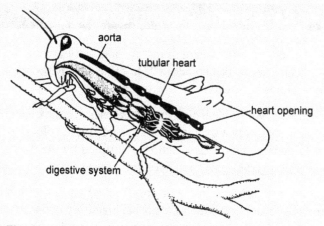

Figure 6.6 "Open" circulatory system of the grasshopper

Table 6.1
Summary of types of circulation present in certain organisms

Organism	System of Circulation
Protozoan	No system. Cyclosis, endoplasmic reticulum and diffusion distribute materials.
Hydra	No system. Same as protozoan.
Earthworm	Closed circulatory system. Has "hearts" (aortic-loop), arteries, veins, capillaries.
Grasshopper	Open circulatory system. Has heart, aorta, sinuses.
Human	Closed circulatory system. Has heart, arteries, veins, capillaries, and lymph vessels.

Problem Solving Example:

The circulatory system of insects does not function in gas exchange. What is its function? Describe the circulatory and respiratory systems in insects.

Insects, which have high metabolic rates, need oxygen in large
amounts. However, insects do not rely on the blood to supply
oxygen to their tissues. This function is fulfilled by the tracheal system. The blood serves only to deliver nutrients and remove wastes.

The insect heart is a muscular dorsal tube that lies within a pericardial sinus. The pericardial sinus is separated by connective tissue
from the perivisceral sinus, which is the hemocoel surrounding the
other internal structures. Usually an insect's only vessel besides the
heart is an anterior aorta. Blood flow is normally posterior to anterior
in the heart and anterior to posterior in the perivisceral sinus. Blood
from the perivisceral sinus drains into the pericardial sinus. The heart
is pierced by a series of openings or ostia, which are regulated by
valves so that blood only flows in one direction. When the heart contracts, the ostia close, and blood is pumped forward. When the heart
relaxes, the ostia open, and blood from the pericardial sinus is drawn
into the heart through the ostia. After leaving the heart and aorta, the
blood fills the spaces between the internal organs, bathing them directly. The rate of blood flow is regulated by the motion of the muscles
of the body wall or the gut.

A respiratory system delivers oxygen directly to the tissues in the
insect. A pair of openings called spiracles is present on the first seven
or eight abdominal segments and on the last one or two thoracic segments. Usually, the spiracle is provided with a valve for closing and
with a filtering apparatus (composed of bristles) to prevent entrance
of dust and parasites.

The organization of the internal tracheal system is quite variable,
but usually a pair of longitudinal trunks with cross connections is
found. Larger tracheae are supported by thickened rings of cuticle,
called taenidia. The tracheae are widened in various places to form

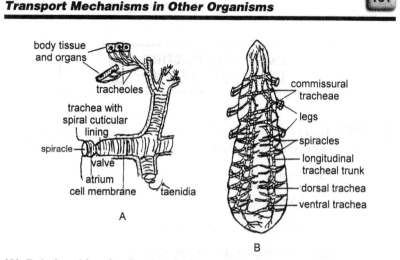

(A) Relationship of spiracle, tracheae, taenidia (chitinous bands that strengthen the tracheae), and tracheoles (diagrammatic). **(B)** Generalized arrangement of insect tracheal system (diagrammatic). Air sacs and tracheoles not shown.

internal air sacs. The air sacs have no taenidia and are sensitive to ventilation pressures (see above). The tracheae branch to form smaller and smaller subdivisions, the smallest being the tracheoles. The smallest tracheoles are in direct contact with the tissues and are filled with fluid at their tips. This is where gas exchange takes place.

Within the tracheae, gas transport is brought about by diffusion, ventilation pressures, or both. Ventilation pressure gradients result from body movements. Body movements causing compression of the air sacs and certain elastic tracheae force air out; those causing expansion of the body wall result in air rushing into the tracheal system. In some insects, the opening and closing of spiracles is coordinated with body movements. Grasshoppers, for example, draw air into the first four pairs of spiracles as the abdomen expands, and expel air through the last six pairs of spiracles as the abdomen contracts.

Quiz: Gas Exchange – Circulatory Systems

1. What type of leaf structures and environmental conditions promote gas exchange in plants?

    (A) Cortex – heat

    (B) Cortex – cold

    (C) Mesophyll – high humidity

    (D) Stomata – normal temperatures

2. The process by which gases are exchanged in the alveoli of our lungs is called

    (A) external respiration.

    (B) indirect respiration.

    (C) internal respiration.

    (D) direct respiration.

3. The only artery in the human body which carries deoxygenated blood is the

    (A) pulmonary artery.

    (B) right coronary artery.

    (C) left coronary artery.

    (D) carotid artery.

4. Which of the following is not a true statement?

    (A) Blood enters the heart through the superior (anterior) vena cava or through the inferior (posterior) vena cava.

    (B) The pulmonary artery carries oxygenated blood.

    (C) Deoxygenated blood first enters the left atrium of the heart.

    (D) The systemic circulation contains oxygenated blood.

5. Functions of the circulatory system include

    (A) delivery of oxygen.

    (B) removal of CO_2.

    (C) transport of hormones.

    (D) All of the above

6. Which of the following correctly shows the path of the blood in the blood vessels?

    (A) Arterioles → capillaries → arteries → veins → venules

    (B) Arteries → arterioles → capillaries → venules → veins

    (C) Capillaries → arterioles → arteries → veins → venules

    (D) Venules → capillaries → veins → arteries → venules

7. Which is not a function of the blood?

    (A) Transport of nutrients and oxygen to the cells

    (B) Removal of carbon dioxide and other wastes

    (C) Transport of hormones to target organs

    (D) All of the above are functions of the blood.

8. The blood plasma contains a number of inorganic cations and anions. If NaCl increases in concentration in the plasma, thus increasing the number of ions, what will be the result?

    (A) Blood pressure will decrease.

    (B) Blood pressure will increase.

    (C) Blood pressure will remain unchanged.

    (D) Blood pressure will sharply increase and then decrease.

9. During the clotting process, fibrinogen is converted into

    (A) thromboplastin.

    (B) fibrin.

    (C) platelets.

    (D) prothrombin.

10. The pumping chambers of the heart are called

    (A) atria.

    (B) ventricles.

    (C) pacemakers.

    (D) cardiac muscles.

ANSWER KEY

1.	(D)	6.	(B)
2.	(B)	7.	(D)
3.	(A)	8.	(B)
4.	(B)	9.	(B)
5.	(D)	10.	(B)

CHAPTER 7

Excretion and Homeostasis

7.1 Excretion in Plants

Plants lack specific organs of excretion and reuse most of the wastes that they produce.

A) Catabolism in plants is usually much lower than in animals; therefore, metabolic wastes accumulate at a slower rate.

B) Green plants use much of the waste products produced by catabolism in anabolic processes.

Problem Solving Example:

  How do higher plants get rid of their waste products?

Since green plants undergo both photosynthesis and respiration, the products of one process may become the raw materials of the other and vice versa. Thus oxygen, the product of photosynthesis, is utilized in cellular respiration while the products of respiration, carbon dioxide and water, are used in photosynthesis. These products may also diffuse out through the pores of the leaves, depending on the dominance of either process at a particular time. For example, during the night, when respiration predominates, carbon dioxide and water vapor escape from the pores of the leaf surfaces.

The amount of nitrogenous wastes in plants is small compared to that in animals, and can be eliminated by diffusion either as ammonia gas through the pores of the leaves or as nitrogen-containing salts through the membrane of the root cells into the soil. Other wastes such as oxalic acid accumulate in the cells of the leaves and are eliminated from the plant when the leaves are shed.

7.2 Excretion in Humans

The organs of excretion which remove metabolic wastes from a cell or organ in humans include the skin, lungs, liver, kidneys, and large intestine.

A) **Skin** – The skin functions as an organ of excretion as well as protection against injury and regulation of body fluids. The sweat glands of the skin remove water, mineral salts, and urea from the blood.

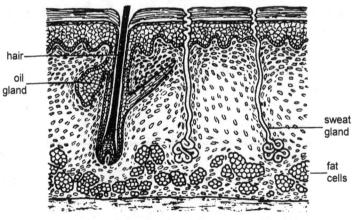

Figure 7.1 Section of the human skin

B) **Lungs** – The lungs excrete water and carbon dioxide.

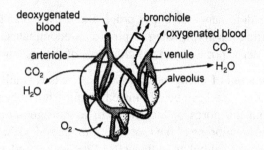

Figure 7.2 Representation of alveoli where carbon dioxide and water are eliminated

C) **Liver** – The liver is the gland that secretes bile for the emulsification of fats. It is considered an organ of excretion because:

1) it removes old red blood cells from the bloodstream.

2) amino acids in excess of the body's anabolic needs are deaminated by the liver.

3) it converts ammonia in the urea cycle.

4) it breaks down toxic substances in the blood.

D) **Urinary System** – The urinary system of humans consists of a pair of kidneys, a pair of ureters, the urinary bladder, and the urethra.

1) **Kidneys** – The kidneys are located against the dorsal body wall just below the diaphragm. They are composed of three distinct regions: the cortex, medulla, and pelvis. The capillaries and tubules in the kidneys form nephrons which remove metabolic wastes from the blood. Blood reaches the kidneys via a right and left renal artery and leaves via right and left renal veins.

a) **Nephron** – structural and functional unit of the kidney which manufactures urine

b) **Glomerulus** – network of capillaries which constitutes part of a single nephron

c) **Bowman's Capsule** – double-walled chamber which surrounds the glomerulus

d) **Proximal Tubule** – segment of the nephron tubule which reclaims sodium ions, glucose, and amino acids

e) **Loop of Henle** – The proximal tubule leads into the loop of Henle where sodium ions are actively transported out of the segment.

f) **Distal Tubule** – Additional sodium can be pumped out by the distal tubule because it is variably permeable to water.

g) **Collecting Tubule** – receives urine from smaller tubules

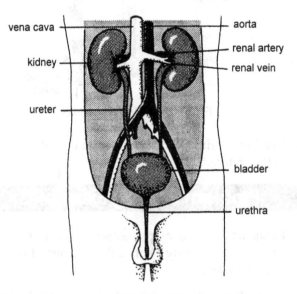

Figure 7.3 The human excretory system

2) **Ureters** – Urine flows down from the kidney to the bladder by means of the ureter.

3) **Bladder** – The bladder is a hollow, muscular organ which is capable of expanding when urine flows into it.

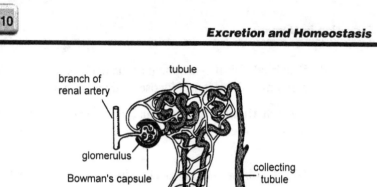
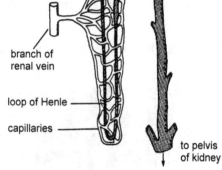

Figure 7.4 A single nephron

4) **Urethra** – Urine flows to the outside from the bladder by way of the urethra.

Problem Solving Examples:

 In humans, the kidney performs the bulk of the excretion of wastes from the body. Outline the structure of the human kidney and urinary system.

 Located on each posterior side of the human body just below the level of the stomach are the bean-shaped kidneys. Each kidney consists of three parts: an outer layer called the cortex, an inner layer called the medulla, and a sac-like chamber called the pelvis. The functional unit of a kidney is the nephron; there are about a million nephrons per kidney. The mechanisms by which the kidneys perform their functions depend on both the physical and physiological relationships between these two components of the nephron.

Throughout its course, the kidney tubule is composed of a single layer of epithelial cells which differs in structure and function from one portion of the tubule to another. The blind end of the tubule is Bowman's capsule, a sac embedded in the cortex and lined with thin epithelial cells. The curved side of Bowman's capsule is in intimate contact with the glomerulus, a compact tuft of branching blood capillaries, while the other opens into the first portion of the tubular system called the proximal convoluted tubule. The proximal convoluted tubule leads to a portion of the tubule known as the loop of Henle. This hairpin loop consists of a descending and an ascending limb, both of which extend into the medulla. Following the loop, the tubule once more becomes coiled as the distal convoluted tubule. Finally, the tubule runs a straight course as the collecting duct. From the glomerulus to the beginning of the collecting duct, each of the million or so nephrons is completely separate from its neighbors. However, the collecting ducts from separate nephrons join to form common ducts, which in turn join to form even longer ducts, which finally empty into the base of each kidney. The renal pelvis is continuous with the ureter, which empties into the urinary bladder where urine is temporarily stored. The urine remains unchanged in the bladder, and when eventually excreted, has the same composition as when it left the collecting ducts.

Blood enters the kidney through the renal artery, which upon reaching the kidney divides into smaller and smaller branches. Each small artery gives off a series of arterioles, each of which leads to a glomerulus. The glomerulus protrudes into the cup of Bowman's capsule and is completely surrounded by the epithelial lining of the capsule. The functional significance of this anatomical arrangement is that blood in the capillaries of the glomerulus is separated from the space within Bowman's capsule only by two extremely thin layers. This thin barrier permits the filtration of plasma (the non-cellular blood fraction) from the capillaries into Bowman's capsule.

 The terms defecation, excretion, and secretion are sometimes confused. What is meant by these terms?

 Defecation refers to the elimination of wastes and undigested food, collectively called feces, from the anus. Undigested food materials have never entered any of the body cells and have not taken part in cellular metabolism; hence they are not metabolic wastes. Excretion refers to the removal of metabolic wastes from the cells and bloodstream. The excretion of wastes by the kidneys involves an expenditure of energy by the cells of the kidney, but the act of defecation requires no such effort by the cells lining the large intestine. Secretion refers to the release from a cell of some substance which is utilized either locally or elsewhere in some body processes; for example, the salivary glands secrete saliva, which is used in the mouth in the first step of chemical digestion. Secretion also involves cellular activity and requires the expenditure of energy by the secreting cell.

 How does the body control the amount of water excreted?

Factors such as blood volume and glomerular capillary pressure act to regulate the amount of fluid initially absorbed by the kidney. The volume of urine excreted, however, is ultimately controlled by the permeability of the walls of the distal convoluted tubules and collecting ducts to water. This permeability can be varied, and is regulated by a hormone known as vasopressin or antidiuretic hormone (ADH). In the absence of ADH, the water permeability of the distal convoluted tubule (DCT) and collecting tubule is very low, and the final urine volume is correspondingly high. In the presence of ADH, water permeability is high and the final urine volume is small. ADH has no effect on sodium absorption, but it regulates the ability of water to osmotically follow ionic absorption.

The release of ADH is regulated by receptors in the hypothalamus. Increased plasma osmolarity causes increased secretion of ADH. Decreased osmolarity (low blood volume) leads to decreased secretion of ADH. Blood volume also influences ADH secretion. Increased blood volume stimulates the decrease of ADH secretion.

Let us look at the interaction of these regulatory mechanisms in a specific situation. If a man drinks an excessive amount of water but does not increase his sodium intake, the most logical way to maintain optimal osmotic concentrations in the body would be to excrete the excess water without altering normal salt excretion. Intake of the excess water results in an increase in extracellular and blood fluid volumes. This has a two-fold effect: First, the osmotic concentration of the blood decreases. This stimulates the osmoreceptors to cause decreased secretion of ADH. Second, the arterial baroreceptors are stimulated and send impulses to the hypothalamus, where ADH release is inhibited. The permeability of the collecting tubules to water is lowered, and consequently more water is excreted.

7.3   Excretion in Other Organisms

A) **Protozoans** – Elimination of metabolic wastes occurs by diffusion through the plasma membrane. The major waste products include ammonia, carbon dioxide, mineral salts, and water. Some have contractile vacuoles for elimination.

B) **Hydra** – The metabolic wastes of the hydra are excreted by simple diffusion. The major wastes of the hydra include ammonia, water, carbon dioxide, and salts.

C) **Grasshopper** – The excretory system of the grasshopper is made up of Malpighian tubules. Wastes such as water, salts, and dissolved nitrogenous compounds diffuse into the blood in the body cavity. The Malpighian tubules absorb these wastes. Water present in the Malpighian tubules is reabsorbed into the blood. Any remaining waste passes into the intestine where it is eliminated. Salts, uric acid, and small quantities of water are the major metabolic wastes.

D) **Earthworm** – The nephridia are the major excretory structures of the earthworm. They are needed for the elimination of water, urea, ammonia, and mineral salts. Each nephridium consists of a nephrostome which lies within the body cavity. This body cavity is filled with fluid which enters the nephrostome and passes down a long tubule. As it travels down this tubule, useful materials are reclaimed by cells which line the tubule. Useless materials leave the earthworm by the nephridiopores, which are openings to the outside.

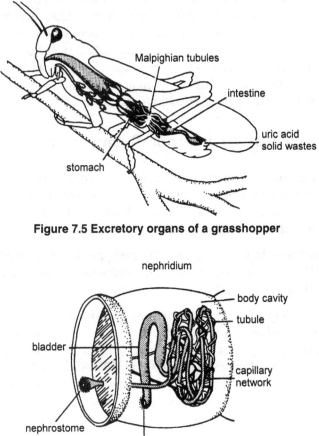

Figure 7.5 Excretory organs of a grasshopper

nephridium

bladder

body cavity

tubule

capillary network

nephrostome

nephridiopore

Figure 7.6 Excretory system of the earthworm

Table 7.1 Summary of Excretion

Organism	Major Organs for Excretion	Major Waste Products
Protozoans	Plasma membrane	water, ammonia, carbon dioxide, salts
Hydra	Plasma membrane	water, ammonia, carbon dioxide, salts
Grasshopper	Malpighian tubules and intestine	uric acid, salts, very little water
Earthworm	Nephridia	urea, water, salts, ammonia
Human	Lungs, kidneys, skin, liver, large intestine	bile, salts, carbon dioxide

Problem Solving Example:

  What is the function of the contractile vacuole?

Unicellular and simple multicellular organisms lack special excretory structures for the elimination of nitrogenous wastes. In these organisms, wastes are simply excreted across the general cell membranes. Some protozoans do, however, have a special excretory organelle called the contractile vacuole. It appears that this organelle eliminates water from the cell but not nitrogenous wastes. These organelles are more common in fresh water protozoans than in marine forms. This is because, in fresh water forms, the concentration of solutes is greater in the cytoplasm than in the surrounding water, and water passively flows into the cell through the cell membrane. The fresh water protozoan is said to live in a hypotonic environment. This inflow of water would cause a fatal bloating if the water were not removed by the contractile vacuole. The vacuole swells and shrinks in a steady cycle, slowly ballooning as water collects in it, then rapidly contracting as it expels its contents, and then slowly ballooning again. The exact process of how the cell pumps water out of the vacuole is still unclear, but it is believed that the process is an energy consuming one.

7.4 Homeostasis

Homeostasis is the automatic maintenance of a steady state within the bodies of all organisms. It is the tendency of organisms to maintain constant conditions of their internal environment by responding to both internal and external changes. The kidney, for example, maintains a constant environment by excreting certain substances and conserving others.

Problem Solving Example:

 There are three processes which together enable the kidney to remove wastes while conserving the useful components of the blood. What are these processes and where do they occur?

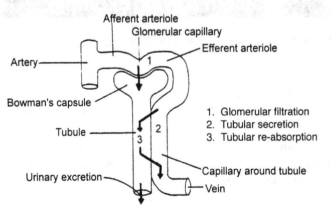

The three basic components of renal function

 Blood flowing to the kidneys first undergoes glomerular filtration. This occurs at the junction of the glomerular capillaries and the wall of Bowman's capsule. The blood plasma is filtered as it passes through the capillaries, which are freely permeable to water and solutes of small molecular dimension yet relatively impermeable to large molecules, especially the plasma proteins. Water, salts, glucose, urea, and other small species pass from the blood into the cavity of Bowman's capsule to become the glomerular filtrate.

It has been demonstrated that the filtrate in Bowman's capsule contains virtually no protein and that all low weight crystalloids (glucose, protons, chloride ions, etc.) are present in the same concentrations as in plasma.

If it were not for the process of tubular reabsorption, the composition of the urine would be identical to that of the glomerular filtrate. This would be extremely wasteful, since a great deal of water, glucose, amino acids and other useful substances present in the filtrate would be lost. Tubular reabsorption is the transfer of material from the tubular lumen back to the blood through the walls of the capillary network. The principal portion of the tubule involved in reabsorption is the proximal convoluted tubule. As the filtrate passes through the tubule, the epithelial cells reabsorb much of the water and virtually all the glucose, amino acids, and other substances useful to the body. The cells then secrete these back into the bloodstream. The secretion of these substances into the blood is accomplished against a concentration gradient, and is thus an energy consuming process – one utilizing ATP. The rates at which substances are reabsorbed, and therefore the rates at which wastes are excreted (because what is not reabsorbed is eliminated) are constantly subjected to physiological control. The ability to vary the excretion of water, sodium, potassium, hydrogen, calcium, and phosphate ions, and many other substances is the essence of the kidney's ability to regulate the internal environment. Reabsorption also occurs in the distal convoluted tubules, where sodium is actively reabsorbed under the influence of aldosterone, a hormone secreted by the adrenal cortex. When this occurs, chloride passively follows due to an electrical gradient; water is also reabsorbed because of an osmotic gradient established by the reabsorption of sodium and chloride. In addition, reabsorption of water takes place in the distal convoluted tubule and collecting duct, stimulated by the posterior pituitary hormone vasopressin, also known as antidiuretic hormone (ADH). ADH increases the permeability of the distal convoluted tubule and collecting duct to water, allowing water to leave the lumen of the nephron, concentrating the urine.

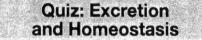
Quiz: Excretion and Homeostasis

1. The functional unit of the kidney is called a

    (A) tubule.

    (B) neuron.

    (C) urethra.

    (D) nephron.

2. In humans, urine is temporarily stored in the _____ before it is voided.

    (A) intestine

    (B) liver

    (C) bladder

    (D) nephron

3. The thin barrier at Bowman's capsule allows for the filtration of

    (A) whole blood.

    (B) plasma.

    (C) oxygen.

    (D) electrolytes only.

4. All of the following are examples of actively transported solutes in the kidney EXCEPT

    (A) urea.

    (B) glucose.

    (C) phosphate.

    (D) urine.

5. If someone suffers from kidney failure, it might be traced to a breakdown in

    (A) secretion.

    (B) filtration.

    (C) reabsorption.

    (D) Any of the above

6. The removal of metabolic waste products that become toxic when they accumulate is known as

    (A) secretion.

    (B) excretion.

    (C) osmosis.

    (D) respiration.

7. The kidney can do all of the following EXCEPT

    (A) remove metabolic wastes.

    (B) help activate vitamin C.

    (C) help regulate blood pressure.

    (D) help stimulate production of red blood cells.

8. Which of the following is not part of the nephron in the human kidney?

    (A) Proximal convoluted tubule

    (B) Loop of Henle

    (C) Distal convoluted tubule

    (D) Major calyx

9. Which organ is not part of the human urinary tract?

    (A) Ureter

    (B) Urethra

    (C) Uterus

    (D) Kidney

10. All of the following are examples of homeostatic mechanisms EXCEPT

    (A) after exercising vigorously, you perspire and the evaporation of water from the skin lowers body temperature.

    (B) phagocytic cells in a rabbit detect and eat bacteria.

    (C) a frog deposits its eggs in a pond.

    (D) shivering occurs outdoors when the air temperature is low.

ANSWER KEY

1. (D)
2. (C)
3. (B)
4. (A)
5. (D)

6. (B)
7. (B)
8. (D)
9. (C)
10. (C)

CHAPTER 8

Hormonal Control in Animals and Plants

8.1 Hormones

A hormone is a chemical substance secreted by specific cells in one area of the body to be used in another area known as the target organ. The target organ responds by being either stimulated or inhibited.

Endocrine glands, or ductless glands, secrete their products directly into the capillaries and become part of the bloodstream.

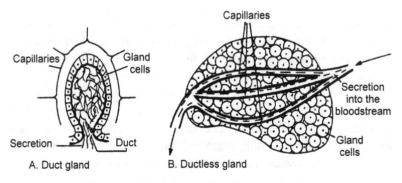

Figure 8.1 Exocrine and endocrine glands

Exocrine glands secrete hormones which pass through ducts, to reach the place where they function. They are also called duct glands.

8.2 The Mechanism and Action of Hormones

A) Only small concentrations are needed for a hormone to be effective.

B) The hormones secreted by an endocrine gland are influenced by certain substances present in the blood and by nerve impulses from the autonomic nervous system.

C) The functioning of one endocrine gland will also affect the functioning of other glands in the body.

D) Current theories include the major principle that hormones combine with specific receptors when the hormone enters the cell.

Problem Solving Example:

 Distinguish between a hormone and an enzyme.

 A hormone is an organic substance synthesized by a specific organ or tissue and secreted into the blood, which carries it to other sites in the body where the hormonal actions are exerted.

Enzymes are organic compounds that function as biochemical catalysts produced by living organisms, and are primarily constituted of protein. Unlike hormones, which may travel considerable distances to target sites, enzymes generally are used at, or very near to, their site of production. For example, pancreatic enzymes exert their effect in the duodenum, which is near the pancreas. Protein hormones that act through the adenyl cyclase system generally elicit their responses by activating certain enzymes in their target cells. Whereas enzymes are always made up of protein, hormones may consist of steroids. Also, enzymes are not released into the blood as are hormones.

8.3 The Human Endocrine System

The major glands of the human endocrine system include the thyroid gland, parathyroid glands, pituitary gland, pancreas, adrenal glands, pineal gland, thymus gland, and the sex glands.

A) **Thyroid Gland** – The thyroid gland is a two-lobed structure located in the neck. It is responsible for the secretion of the hormone thyroxin and for iodide absorption. Thyroxin increases the rate of cellular oxidation and influences growth and development of the body.

B) **Parathyroid Glands** – The parathyroid glands are located in back of the thyroid gland. They secrete the hormone parathormone which is responsible for regulating the amount of calcium and phosphate salts in the blood.

C) **Pituitary Gland** – The pituitary gland is located at the base of the brain. It consists of three lobes: the anterior lobe; the intermediate lobe, which is only a vestige in adulthood; and the posterior lobe.

1) Hormones of the anterior lobe

   a) **Growth Hormone** – stimulates growth of bones

   b) **Thyroid-stimulating Hormone (TSH)** – stimulates the thyroid gland to produce thyroxin

   c) **Prolactin** – regulates development of the mammary glands of a pregnant female and stimulates secretion of milk in a woman after childbirth

   d) **Adrenocorticotropic Hormone (ACTH)** – stimulates the secretion of hormones by the cortex of the adrenal glands

   e) **Follicle-stimulating Hormone (FSH)** – this hormone acts upon the gonads, or sex organs

   f) **Luteinizing Hormone (LH)** – in the male, LH causes the cells in the testes to secrete androgens. In females, LH causes the follicle in an ovary to change into the corpus luteum.

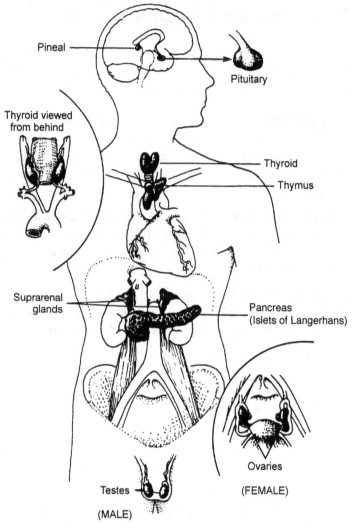

Figure 8.2 The human endocrine system

2) Hormones of the intermediate lobe

This lobe secretes a hormone that has no known effects in humans.

3) Hormones of the posterior lobe

    a) **Vasopressin (ADH)** – This hormone causes the muscular walls of the arterioles to contract, thus increasing blood pressure. It regulates the amount of water reabsorbed by the nephrons in the kidney.

    b) **Oxytocin** – This hormone stimulates the muscle of the walls of the uterus to contract during childbirth. It induces labor.

Problem Solving Example:

Q The parathyroid glands in humans are small, pea-like organs. They are usually four in number and are located on the surface of the thyroid. What is the function of these glands?

A The parathyroids were long thought to be part of the thyroid or to be functionally associated with it. Now, however, we know that their close proximity to the thyroid is misleading; both developmentally and functionally, they are totally distinct from the thyroid.

The parathyroid hormone, called parathormone, regulates the calcium-phosphate balance between the blood and other tissues. Production of this hormone is directly controlled by the calcium concentration of the extracellular fluid bathing the cells of these glands. Parathormone exerts at least four effects: (1) it increases gastrointestinal absorption of calcium; (2) it increases the movement of calcium and phosphate from bone into extracellular fluid; (3) it increases reabsorption of calcium by the renal tubules; (4) it reduces the reabsorption of phosphate by the renal tubules.

The first three effects result in a higher extracellular calcium concentration. The adaptive value of the fourth is to prevent the formation of kidney stones.

8.4 The Pancreas

The pancreas is both an endocrine and an exocrine gland. As an endocrine gland, the islets of Langerhans, scattered through the pancreas, secrete insulin and glucagon.

A) **Insulin** – acts to lower the level of glucose in the bloodstream. Glucose is converted to glycogen.

B) **Glucagon** – increases the level of glucose in the blood by helping to change liver glycogen into glucose.

Problem Solving Example:

Q The pancreas is a mixed gland having both endocrine and exocrine functions. The exocrine portion secretes digestive enzymes into the duodenum via the pancreatic duct. The endocrine portion secretes two hormones (insulin and glucagon) into the blood. What are the effects of these two hormones?

A Insulin is a hormone which acts directly or indirectly on most tissues of the body, with the exception of the brain. The most important action of insulin is the stimulation of the uptake of glucose by many tissues, particularly liver muscle and fat. The uptake of glucose by the cells decreases blood glucose levels and increases the availability of glucose for those cellular reactions in which glucose participates.

As stated, insulin stimulates glycogen synthesis, increasing the activity of the enzyme which catalyzes the reaction. The net protein synthesis is also increased by insulin, which stimulates the active membrane transport of amino acids, particularly into muscle cells. Insulin also has effects on other liver enzymes, but the precise mechanisms by which insulin induces these changes are not fully understood.

Insulin secretion is directly controlled by the glucose concentration of the blood flowing through the pancreas. This is a simple system which requires no participation of nerves or other hormones.

Insulin and glucagon are secreted by the Islets of Langerhans. Glucagon is a hormone which has the following major effects: it increases glycogen breakdown, thereby raising the plasma glucose level, and it increases hepatic synthesis of glucose. The glucagon-secreting cells in the pancreas respond to changes in the concentration of glucose in the blood flowing through the pancreas; no other nerves or hormones are involved.

It should be noted that glucagon has the opposite effects as insulin. Glucagon elevates plasma glucose whereas insulin stimulates its uptake and thereby reduces plasma glucose levels.

8.5 Adrenal Glands

The two adrenal glands are located on top of each kidney. They are composed of two regions: the adrenal cortex and the adrenal medulla.

A) Hormones of the Adrenal Cortex

1) **Cortisones** – regulate the change of amino acids and fatty acids into glucose. They also help to suppress reactions that lead to the inflammation of injured parts.

2) **Cortins** – regulate the use of sodium and calcium salts by the body cells.

3) **Sex Hormones** – they are similar in chemical composition to hormones secreted by sex glands.

B) Hormones of the Adrenal Medulla

1) **Epinephrine** – this hormone is responsible for the release of glucose from the liver, the relaxation of the smooth muscles of the bronchioles, dilation of the pupils of the eye, a reduction in the clotting time of blood, and an increase in the heart beat rate, blood pressure, and respiration rate.

2) **Norepinephrine** – this hormone is responsible for the constriction of blood vessels.

Problem Solving Example:

Q The two adrenal glands lie very close to the kidneys. Each adrenal gland in mammals is actually a double gland, composed of an inner core-like medulla and an outer cortex. Each of these is functionally unrelated. Outline the function of the adrenal medulla.

A The adrenal medulla secretes two hormones, adrenaline (epinephrine) and noradrenalin (norepinephrine, NE), whose functions are very similar but not identical.

Epinephrine promotes several responses, all of which are helpful in coping with emergencies: the blood pressure rises, the heart rate increases, the glucose content of the blood rises because of glycogen breakdown, the spleen contracts and squeezes out a reserve supply of blood, the clotting time of blood is decreased, the pupils dilate, the blood flow to skeletal muscle increases, the blood supply to intestinal smooth muscle decreases, and hairs become erect. These adrenal functions, which mobilize the resources of the body in emergencies, have been called the fight-or-flight response. Norepinephrine stimulates reactions similar to those produced by epinephrine, but is less effective in the conversion of glycogen into glucose.

8.6   Pineal Gland

The pineal gland is attached to the brain above the cerebellum. It is responsible for the production of melatonin, a hormone which some scientists believe helps regulate sleep-wake cycles.

Problem Solving Example:

Q The pineal gland is a lobe on the upper portion of the forebrain. What is its function?

A The pineal gland, a pea-sized structure attached above the cerebellum, secretes a hormone called melatonin as the eyes register darkness. At night, melatonin is produced to aid in the sleep-wake cycles of the body. The amount of melatonin produced lessens with age, which is why some scientists believe older people have a harder time sleeping than young people.

8.7   Thymus Gland

The thymus gland is located under the breastbone. Although there is no convincing evidence for its role in the human adult, it does secrete thymosin hormone in infants which stimulates the formation of an antibody system.

Problem Solving Example:

Q   The thymus gland is a two-lobed, glandular-appearing structure located in the upper region of the chest just behind the sternum. What are the two principal functions that have been attributed to this gland?

A   It is thought that one of the functions of the thymus is to provide the initial supply of lymphocytes. These primary cells then give rise to descendent lines of lymphocytes, making further release from the thymus unnecessary. This first function of the thymus is non-endocrine in nature.

The second function attributed to the thymus is the release of the hormone thymosin which stimulates the differentiation of incipient plasma cells in the lymphoid tissues. The cells then develop into functional plasma cells, capable of producing antibodies when stimulated by the appropriate antigens.

8.8   Sex Glands

These glands include the testes of the male and the ovaries of the female.

A)   **Testes** – Luteinizing hormone stimulates specific cells of the testes to secrete androgens. Testosterone, which controls the development of male secondary sex characteristics, is the principle androgen.

B)   **Ovaries** – Estrogen is secreted from the cells which line the ovarian follicle. This hormone is responsible for the development of female secondary sex characteristics.

Table 8.1 Human Endocrine Glands and Their Functions

Gland	Hormone	Function
Pituitary	Growth hormone	Stimulates growth of skeleton
Anterior lobe	FSH	Stimulates follicle formation in ovaries and sperm formation in testes
	LH	Stimulates formation of corpus luteum in ovaries and secretion of testosterone in testes
	TSH	Stimulates secretion of thyroxin from thyroid gland
	ACTH	Stimulates secretion of cortisone and cortin from adrenal cortex
	Prolactin	Stimulates secretion of milk in mammary glands
Posterior lobe	Vasopressin (ADH)	Controls narrowing of arteries and rate of water absorption in kidney tubules
	Oxytocin	Stimulates contraction of smooth muscle of uterus
Thyroid	Thyroxine	Controls rate of metabolism and physical and mental development
	Calcitonin	Controls calcium metabolism

Table 8.1 (continued)

Parathyroids	Parathormone	Regulates calcium and phosphate levels in blood
Islets of Langerhans		
Beta cells	Insulin	Promotes storage and oxidation of glucose
Alpha cells	Glucagon	Releases glucose into bloodstream
Thymus	Thymus hormone	Stimulates formation of antibody system
Adrenals Cortex	Cortisones	Promote glucose formation from amino acids and fatty acids
	Cortins	Control water and salt balance
	Sex hormones	Influence sexual development
Medulla	Epinephrine (adrenalin) or norepinephrine (noradrenalin)	Releases glucose into bloodstream, increases rate of heartbeat, increases rate of respiration, reduces clotting time, relaxes smooth muscle in air passages
Gonads Ovaries, follicle cells	Estrogen	Controls female secondary sex characteristics
Corpus luteum cells	Progesterone	Helps maintain attachment of embryo to mother
Testes	Testosterone	Controls male secondary sex characteristics

Problem Solving Examples:

Q What would happen if, following ovulation, no corpus luteum was formed?

A After ovulation, a transformation occurs within the remaining follicle in the ovary, giving rise to a yellowish gland-like structure called the corpus luteum. If pregnancy does not occur, the corpus luteum degenerates. If pregnancy does occur, the corpus luteum grows and persists until near the end of the pregnancy. The corpus luteum continues to secrete estrogen. In addition, the corpus luteum cells also secrete progesterone which helps to sustain the placenta for a fetus.

Progesterone functions in preparing the uterus to receive the embryo. This hormone acts on the uterine lining, causing maturation of the complex system of glands in the lining. Estrogen also acts on the uterus to cause a thickening of the lining. The luteal phase follows ovulation, and during this phase progesterone has the primary influence on the development of the uterine lining. This hormone also stimulates breast growth, particularly of glandular tissue.

If a female failed to develop a corpus luteum following ovulation, a hormonal imbalance would result. There would be a lower than normal concentration of estrogen and progesterone in the circulation. As the levels of these two hormones become low, the feedback inhibition of FSH and LH production is removed. The FSH concentration begins to rise in the plasma, and a new menstrual cycle would begin shortly after ovulation. Even worse, if fertilization occurred and the corpus luteum did not form, the uterus would then fail to mature, and implantation of the fertilized egg would be prevented. Even if implantation did occur, subsequent development would be abnormal because of incomplete placental development.

Q What hormones are involved with the changes that occur in a pubescent female and male?

A Puberty begins in the female when the hypothalamus stimulates the anterior pituitary to release increased amounts of FSH (follicle stimulating hormone) and LH (luteinizing hormone). These hormones cause the ovaries to mature, which then begin secreting estrogen and progesterone, the female sex hormones. These hormones, particularly estrogen, are responsible for the development of the female secondary sexual characteristics. These characteristics include the growth of pubic hair, an increase in the size of the uterus and vagina, a broadening of the hips and development of the breasts, a change in voice quality, and the onset of the menstrual cycles.

Before the onset of puberty in the male, no sperm and very little male sex hormone are produced by the testes. The onset of puberty begins, as in the female, when the hypothalamus stimulates the anterior pituitary to release increased amounts of FSH and LH. In the male, FSH stimulates maturation of the seminiferous tubules which produce the sperm. LH is responsible for the maturation of the interstitial cells of the testes. It also induces them to begin secretion of testosterone, the male sex hormone. When enough testosterone accumulates, it brings about the whole spectrum of secondary sexual characteristics normally associated with puberty. These include growth of facial and pubic hair, deepening of the voice, maturation of the seminal vesicles and the prostate gland, broadening of the shoulders, and the development of the muscles.

8.9   Endocrine Abnormalities

A) **Myxedema –** This results when there is a deficiency in the amount of thyroxine that is secreted by the thyroid gland. Myxedema occurs in adults and is characterized by decreased heat production and a low metabolic rate.

B) **Cretinism –** Cretinism occurs when the thyroid gland is defective at birth. It is a type of dwarfism characterized by defective teeth, protruding abdomen, and low mental ability.

C) **Goiter –** An enlarged thyroid gland is known as a goiter. Nontoxic goiter is the result of a lack of iodine in the diet. Toxic goiter results from an overdevelopment of the thyroid gland.

D) **Hyperparathyroidism –** An oversecretion of the parathyroid hormone is responsible for excessive withdrawal of calcium from the bones, causing them to soften and break easily.

E) **Hypoparathyroidism** – Removal of the parathyroids results in a high phosphate concentration in the blood and a low calcium concentration. This produces serious disturbances of muscles and nerves.

F) **Diabetes** – Diabetes results when there is not a sufficient amount of insulin secreted by the beta cells. The insulin cannot regulate the passage of glucose from the blood into the muscles and the liver.

G) **Giantism** – Giantism results when there is an oversecretion of growth hormone during childhood.

H) **Dwarfism** – Dwarfism results when there is an undersecretion of growth hormone during childhood.

I) **Acromegaly** – Acromegaly is the result of an oversecretion of growth hormone in the adult. This results in an overgrowth of parts of the body which can still respond to the hormone such as the ends of the bones of the hands, feet, and face.

Problem Solving Examples:

 Explain why the injection of thyroxine may cure myxedema, and why the oral insulin does not cure diabetes.

 Myxedema is a disease cause by a deficiency of thyroxine, a hormone secreted by the thyroid gland. Ingested thyroxine can be used to treat myxedema because it is absorbed by the gut in an unaltered form. Thus the feeding of thyroid glands is one possible way to cure myxedema.

The hormone produced by the Islets of Langerhans in the pancreas is insulin. Insulin is one of the smallest proteins known, and like other proteins it consists of a sequence of amino acids linked by peptide bonds. Should insulin be ingested and enter the stomach, proteases there would hydrolyze the hormone into individual amino acids, thus destroying the hormone. This is why diabetes cannot be cured by oral administration of insulin. Instead, insulin is usually directly introduced into the blood stream.

Q The thyroid gland is located in the neck and secretes several hormones, the principle one being thyroxine. What functions does it serve in the body? What happens when there is a decreased or increased amount of thyroxine in the body?

A The thyroid gland is a two-lobed gland which manifests a remarkably powerful active transport mechanism for uptaking iodide ions from the blood. As blood flows through the gland, iodide is actively transported into the cells and is combined with amino acids to form thyroxin. The hormone activating thyroxin hormone, produced by the pituitary gland, is known as thyroid-stimulating hormone (TSH). A variety of bodily defects, either dietary, hereditary, or disease-induced, may decrease the amount of thyroxine released into the blood. The most popular of these defects is one which results from dietary iodine deficiency. The thyroid gland enlarges, in the continued presence of TSH from the pituitary, to form a goiter. This is a futile attempt to synthesize thyroid hormones, for iodine levels are low, causing continual stimulation of the thyroid and the inevitable protuberance on the neck. Formerly, the principle source of iodine came from seafood. As a result, goiter was prevalent amongst inland areas far removed from the sea. Today, the incidence of goiter has been drastically reduced by adding iodine to table salt.

Thyroxine serves to stimulate oxidative metabolism in cells; it increases the oxygen consumption and heat production of most body tissues, a notable exception being the brain. Thyroxine is also necessary for normal growth. The absence of thyroxine significantly reduces the ability of growth hormone to stimulate amino acid uptake and RNA synthesis.

If there is an insufficient amount of thyroxine, a condition referred to as hypothyroidism results. Symptoms of hypothyroidism stem from the fact that there is a reduction in the rate of energy-releasing reactions within the body cells. Usually the patient shows puffy skin, sluggishness, and lowered vitality. Hypothyroidism in children, a condition known as cretinism, can result in mental retardation, dwarfism, and permanent sexual immaturity. Sometimes the thyroid gland produces too much thyroxine, a condition known as hyperthyroidism. This condition produces symptoms such as an abnormally high body temperature, profuse sweating, high blood pressure, loss of weight, irritability, and muscular weakness. Hyperthyroidism has been treated by partial removal or by partial radiation destruction of the gland. More recently, several drugs that inhibit thyroid activity have been discovered, and their use is supplanting the former surgical procedures.

8.10 Plant Hormones

A) **Auxins** – These plant growth regulators stimulate the elongation of specific plant cells and inhibit the growth of other plant cells.

B) **Gibberellins** – In some plants, gibberellins are involved in the stimulation of flower formation. They also increase the stem length of some plant species and the size of fruits. Gibberellins also stimulate the germination of seeds.

C) **Cytokinins** – Cytokinins increase the rate of cell division and stimulate the growth of cells in a tissue culture. They also influence the shedding of leaves and fruits, seed germination, and the pattern of branch growth.

Problem Solving Examples:

   What are plant hormones? Discuss the important effects of the auxins.

   Plant hormones, like animal hormones, are organic substances which can produce striking effects on cell metabolism and growth even though present in extremely small amounts. They are produced primarily in actively growing tissues. Plant hormones usually affect parts of the plant body somewhat removed from the site of their production. Movement of plant hormones to target regions of the plant is made possible by the presence of phloem.

Auxins may be regarded as the most important of the plant hormones, since they have the most marked effects in correlating growth and differentiation to result in the normal pattern of development. The differential distribution of auxin in the stem of the plant as it moves down from the apex (where it is produced) causes the plant to elongate and bend toward light. Auxin from seeds induces the maturation of the fruit. The auxins, in addition, determine the growth correlations of the several parts of the plant. They inhibit development of the lateral buds and promote growth of the terminal bud. Finally, auxins control the shedding of leaves, flowers, fruits, and branches from the parent plant.

 Discuss the important effects of gibberellins and cytokinins on plant growth.

 Gibberellins and cytokinins are two major types of plant hormones that have been identified in addition to auxins. They interact with auxins and with each other to regulate biological activities in plants. Gibberellins and cytokinins have dominant roles in controlling the early phases of growth and development; auxins become dominant later in controlling cell elongation.

Gibberellins function to lengthen stems, stimulate seed germination, induce flower formation, and increase the size of fruits in some species of plants. The former two are their most vital functions. Seed germination is examined here: Just before germination, the embryo of a seed secretes a gibberellin which induces the production of an enzyme which hydrolyzes the stored starch for energy. The hormone also activates other enzymes of the seed which break down the materials of the seed coat.

Cytokinins function in stimulating growth of cells and accelerating their rate of division. Cytokinins can also change the structure of plant tissue: at certain concentrations, cytokinins are shown to cause root and shoot formation in plant tissue cultures. Cytokinins are also found to oppose auxins by causing lateral buds to develop and thus modifying apical dominance. In addition, cytokinins have been demonstrated to prevent leaves from yellowing and hence play a role in delaying senescence.

Quiz: Hormonal Control in Animals and Plants

1. Which of the following is not a function of the hormone auxin?

    (A) It causes the plant to bend and elongate toward the light.

    (B) It directs the tissue differentiation in the vascular cambium.

    (C) It stimulates development of lateral buds and inhibits growth of the terminal bud.

    (D) It controls the shedding of leaves, flowers, fruits, and branches.

2. The hormone glucagon

    (A) has the opposite effect on the liver that insulin does.

    (B) converts glycogen into glucose.

    (C) is produced in the beta cells of the pancreas.

    (D) Both (A) and (B)

3. It has long been known that kidney malfunction is commonly associated with high blood pressure (hypertension). What is a possible explanation of this?

    (A) Underproduction of thyroxin

    (B) Overproduction of renin

    (C) Relaxation of the smooth muscles in the walls of the blood vessels

(D) Decrease in the level of secretion of aldosterone by the adrenal cortex

4. Goiter is a disease associated with a malfunctioning of the thyroid. Which is NOT a symptom of this disease?

    (A) Swollen neck

    (B) Dry and puffy skin

    (C) Loss of hair

    (D) Rapid heartbeat

5. Which of the following hormones is produced by the thyroid?

    (A) Thyroxin

    (B) Vasopressin

    (C) Calcitonin

    (D) Both (A) and (C)

6. Which of the following hormones are secreted by the adrenal medulla?

    (A) Cortical sex hormones and aldosterone

    (B) Aldosterone and adrenalin

    (C) Noradrenalin and adrenalin

    (D) Cortisone and noradrenalin

7. The "Fight-or-Flight" reaction is characterized by the increase of glucose levels and oxygen content to the skeleton and muscles, with a decrease in glucose level and oxygen content to the digestive system. What hormone(s) is/are responsible for this reaction?

    (A) Adrenalin

    (B) Noradrenalin

    (C) Cortisone

    (D) Both (A) and (B)

8. Which of the following hormones should be associated with the bearded lady in the circus?

    (A) Glucocorticoids

    (B) Cortical sex hormones

    (C) Testosterone

    (D) Progesterone

9. Vasopressin and oxytocin are two hormones associated with the

    (A) posterior pituitary.

    (B) anterior pituitary.

    (C) pineal gland.

    (D) thyroid.

10. Which of the following hormones influence(s) the reabsorption of water in the kidney?

    (A) ADH

    (B) Aldosterone

    (C) Insulin

    (D) Both (A) and (B)

ANSWER KEY

1.	(C)	6.	(C)
2.	(D)	7.	(D)
3.	(B)	8.	(B)
4.	(D)	9.	(A)
5.	(D)	10.	(D)

CHAPTER 9

Stimulus Receptors in Animals

9.1 Components of Nervous Coordination

A) **Receptor** – This is a structure that must have the ability to detect a change in the environment and initiate a signal in the nerve cell.

B) **Conductors** – These are the structures which conduct impulses.

C) **Effectors** – These are the structures that respond to the stimuli, which is information received by receptors.

Problem Solving Example:

Q Our eyes are the principal organs involved in seeing. Yet we are able to perform manual acts such as dressing or tying knots when our eyes are closed. Why?

A There are many sense organs (receptors) in the body. Our pair of eyes is just one example. Although our eyes are extremely important to our perception of this world, we would not be totally and helplessly lost if we could no longer use our eyes. For instance, we are still aware of the relation of our body to the environment even if our eyes are closed, i.e., whether we are standing or sitting, where our limbs are, and where one part of our body is in relationship to another. Such perception without the use of our eyes is achieved with a different set of sense organs known as proprioceptors.

Proprioceptors are receptors found in muscles, tendons, and joints, and are sensitive to muscle tension and stretch. They pick up impulses from the movements and positions of muscles and limbs relative to each other and relay them to the cerebellum for coordination. Impulses from the proprioceptors are extremely important in ensuring the coordinated and harmonious contraction of different muscles involved in a single movement. Without them, complicated skillful acts would not be possible.

Proprioceptors have other important functions. They help maintain the sense of balance and give the body general awareness of its environment.

9.2   Photoreceptors

The photoreceptive system includes the sense of sight. The eye is a delicate receptor which is sensitive to light. Tears secreted by tear glands help in keeping the surface of the eyeball moist and dust-free. The eyelashes help to keep dust particles out of the eye, while the eyelids help spread tears over the eyeball's surface.

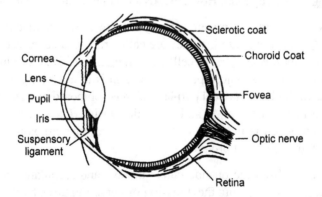

Figure 9.1 The human eye

A) **Sclera** – This is the outer layer of the eyeball. It is white in color except for the cornea which is transparent and admits light into the eye's interior.

B) **Choroid** – The choroid is the middle layer of the eye. Its purpose is to stop the reflection of scattered light within the eye. The choroid coat forms the iris which has a central opening known as the pupil. The pupil permits light to enter the eye. The lens of the eye is located behind the pupil.

C) **Retina** – The retina is the innermost layer of the eye. It contains two types of receptor nerve cells known as the rods and cones which are the actual visual receptors.

   1) **Rods** – These are sensitive to light and are basically used for vision in dim light.

   2) **Cones** – These function as bright-light fine vision color receptors. Both rods and cones are connected to a network of nerve cells which forms the optic nerve. The optic nerve is connected to the brain.

Problem Solving Example:

 Discuss the mechanism by which the photoreceptors are stimulated by light. How are rods and cones distributed in the retina?

 Photoreceptors are sensory cells that are sensitive to light. In the human retina, they are called rods and cones according to their shapes. Both types of cells contain light-sensitive molecules called visual pigments, whose primary function is to absorb light. The rods contain rhodopsin (visual purple) which is composed of a chromophore (a variant of vitamin A) and a protein (opsin). The cones contain iodopsin which is made up of the same chromophore as in rhodopsin but with a different protein.

Light from the outside enters the eye and stimulates the rods or cones, thus triggering the emission of nerve impulses by the receptor cells. Light does not directly provide the necessary energy to set off the impulse. The energy comes from the chemical bonds in the rhodopsin or iodopsin molecule.

When light strikes the visual pigments, it acts upon the chromophore, which then splits away from the opsin because light changes the molecular configuration of the chromophore in such a way that it no longer can bind to the opsin. Simultaneously, impulses are triggered, and these travel to the brain (via the optic nerve) where they are interpreted.

Rhodopsin is sensitive to a very small amount of light. Rod cells, which contain this kind of visual pigment, are used to detect objects in poor illumination such as in night vision. They are not responsible for color vision, but are important in the perception of shades of gray and brightness. Their acuity – ability to distinguish one point in space from another nearby point – is very poor. Rods are most numerous in the peripheral retina, and are absent from the very center of the retina (the fovea). On the other hand, cones operate only at high levels of illumination and are used for day vision. The primary function of the cones to perceive colors. Their visual acuity is very high, and because they are concentrated in the center of the retina, it is that part which we use for fine, detailed vision.

9.3 Vision Defects

A) **Nearsightedness (Myopia)** – In most cases of nearsightedness, the eyeball is too long; thus, the retina is too far from the lens. The light rays converge in front of the retina and diverge when they reach it. This results in a blurred image. Eyeglasses with concave lenses correct this defect.

B) **Farsightedness (Hypermetropia)** – In most cases of farsightedness, the eyeball is too short, so the retina is too close to the lens. Light rays will strike the retina before they converge which results in a blurred image. Eyeglasses with convex lenses correct farsightedness.

C) **Astigmatism** – In astigmatism, there is an irregularity in the curvature of the cornea or the lens. This causes a blurred image. Astigmatism can be corrected with lenses that correct the irregular curvature of the cornea or the lens.

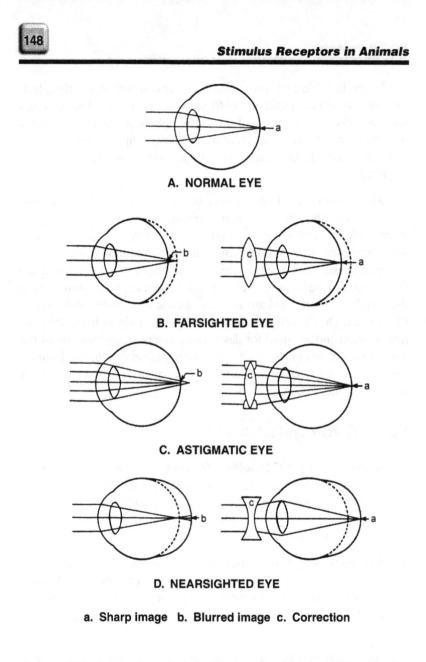

A. NORMAL EYE

B. FARSIGHTED EYE

C. ASTIGMATIC EYE

D. NEARSIGHTED EYE

a. Sharp image   b. Blurred image   c. Correction

Figure 9.2 Illustrations of normal eye and common eye defects with corrective lenses

Problem Solving Example:

Q What are the defects that produce myopia, hypermetropia and astigmatism? What corrective measure can be taken for each defect?

A The most common defects of the human eye are nearsightedness (myopia), which is the inability to see distant objects clearly; farsightedness (hypermetropia), which is the inability to see nearby objects distinctly; and astigmatism, a defect owing to an improperly shaped eyeball or irregularities in the cornea. In the normal eye, the shape of the eyeball is such that the retina is located where light rays converge in the fovea. In a nearsighted eye, the eyeball is too long, and the retina is too far from the lens. The light rays converge at a point in front of the retina, and are again diverging when they reach it, resulting in a blurred image. This defect can be corrected by placing a concave lens in front of the eye, which diverges the light rays before they reach the lens, making it possible for the eye to focus these rays properly on the retina. In a farsighted eye, the eyeball is too short and the retina is thus too close to the lens. Light rays strike the retina before they have converged, again resulting in a blurred image. Convex lenses, when placed in front of the defective eyes, correct for the farsighted condition by causing the light rays to converge farther forward, so that they can come to a focal point on the retina.

Another cause for this visual disorder may be that a myopic eye may have a lens which is too strong – it bends light rays too much. A hypermetropic eye has a lens which is too weak. Whether the cause be from eyeball length or lens strength, a myopic eye will focus the image in front of the retina, while a hypermetropic eye will focus the image behind the retina.

In astigmatism, the cornea is curved unequally in different planes, so that light rays in one plane are focused at a different point from those in another plane. To correct for astigmatism, a cylindrical lens is used which bends light rays going through certain irregular parts of the cornea.

9.4   Chemoreceptors

The senses of taste and smell comprise the chemoreceptive system. These are the receptors of chemicals in the external environment.

A) **Smell** – The nose is the organ in humans which detects odors by means of receptor cells in two olfactory epithelia. An individual receptor consists of a cell with tiny hairs at one end and a nerve cell fiber at the other end. Present theories suggest that active sites on the receptor cells join with specific odor molecules. This combination forms a complex that generates a signal in the receptor cell. The signal then passes through a nerve fiber, which is part of the olfactory nerve, to the brain for interpretation.

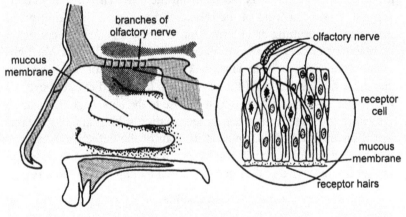

Figure 9.3 Receptors in the nose

B) **Taste** – The taste buds, which are located on the tongue, are the principal receptors of chemical stimuli in the external environment. There are five different types of taste buds which, when stimulated, initiate one primary taste sensation such as sweet, sour, salty, bitter, or umami, a savory taste. Each type of taste receptor cell has its own specific active site which combines with the specific food molecule.

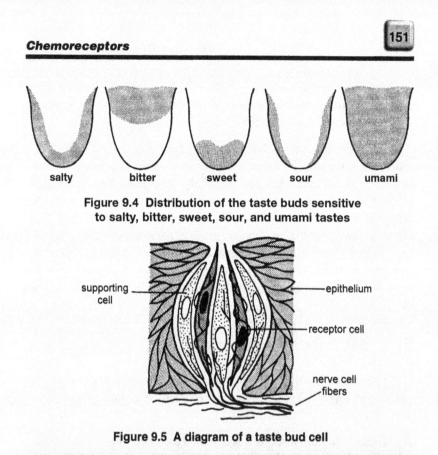

salty   bitter   sweet   sour   umami

Figure 9.4 Distribution of the taste buds sensitive to salty, bitter, sweet, sour, and umami tastes

supporting cell

epithelium

receptor cell

nerve cell fibers

Figure 9.5 A diagram of a taste bud cell

Problem Solving Example:

Q Certain connoisseurs can recognize hundreds of varieties of wine by tasting small samples. How is this possible when there are only five types of taste receptors?

A Taste buds on the tongue and the soft palate are the organs of taste in human beings. Each taste bud contains supportive cells as well as epithelial cells that function as receptors. These epithelial cells have numerous microvilli that are exposed on the tongue surface. Each receptor is innervated by one or more neurons, and when a receptor is excited, it generates impulses in the neurons. There are five basic taste senses: sweet, sour, bitter, salty, and umami. The receptors for each of these basic tastes are concentrated in different regions of the tongue – sweet and salty on the front, bitter on the back, sour on the sides, and umami over the entire surface of the tongue. The sensitivity of these five regions on the tongue to the five different tastes can be demonstrated by placing solutions with various tastes on each region. A dry tongue is insensitive to taste.

Few substances stimulate only one of the five kinds of receptors, but most stimulate two or more types in varying degrees. The common taste sensations we experience daily are created by combinations of the five basic tastes in different relative intensities. Moreover, taste does not depend on the perception of the receptors in the taste buds alone. Olfaction plays an important role in the sense of taste. Together they help us distinguish an enormous number of different tastes. We can now understand how a connoisseur, using a combination of his taste buds and his sense of smell, can recognize hundreds of varieties of wine.

9.5 Mechanoreceptors

The senses involving touch and hearing comprise the category of mechanoreceptors. These receptors are sensitive to mechanical stimuli such as pressure or compression.

A) **Touch** – In humans, touch is detected by receptors near the surface of the skin next to a hair follicle. The Pacinian corpuscle is a receptor found in the skin and in some internal organs. Each Pacinian corpuscle is connected to a sensory neuron. It is a pressure receptor; therefore, any application of pressure will deform the corpuscle.

Proprioceptors are sense receptors distributed throughout skeletal muscle and tendons. Any muscle contraction or stretching will trigger the receptors to initiate nerve impulses.

B) **Hearing** – The organs of hearing and equilibrium are found in the ear. The ear is especially sensitive to sounds of varying frequencies and intensities. The human ear consists of an outer ear, a middle ear, and an inner ear.

   1) **Outer Ear** – The outer ear consists of the pinna, which is an ear flap, and the auditory canal, which provides the passageway for waves to the middle ear. The eardrum separates the outer ear from the middle ear.

2) **Middle Ear** – The middle ear contains three tiny bones known as the hammer, anvil, and stirrup. These bones transmit sound waves across the middle ear cavity. The Eustachian tube connects the middle ear to the pharynx and equalizes the pressure between the outer ear and middle ear.

3) **Inner Ear** – The inner ear consists of the cochlea and three semicircular canals. Lying within the inner chamber of the cochlea is the organ of Corti, which contains the vibration receptors. Nerve impulses, once initiated, travel along the auditory canal to the brain for interpretation as sound.

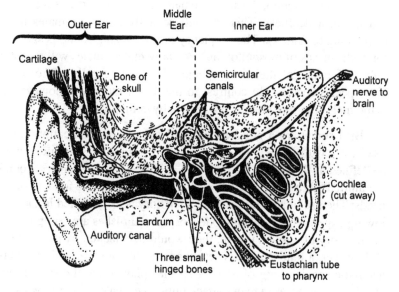

Figure 9.6 The structure of the ear

Problem Solving Example:

 What functions are served by sound communication in animals?

 Communication by sound can convey a variety of messages necessary to the survival and reproduction of the participants. Sound communication is of special interest to us as humans because it serves as the fundamental basis of our language. Other species use forms of auditory communication, although not nearly as complex or sophisticated as human speech. Mate attraction, species recognition, defense, and territorial claim are examples of functions of sound communication in animals.

The female *Aedes* mosquito creates a buzzing sound with her wings, which serves to attract the male mosquito. The sound waves cause the male's antennae to vibrate, stimulating sensory hairs. This allows the male to locate the female and copulate with her. An interesting adaptation for this communication is that in the sexually immature male mosquito, the antennae hairs lie close to the antenna shaft, causing near deafness. This prevents sexually immature males from responding to females. The frequency of the buzzing sound is specific for each species of mosquito; thus, sexually mature males will respond only to females of their own species. The mating calls of frogs are also species-specific. In contrast to the *Aedes* mosquito, however, it is the male frog that attracts the female by calling.

Bird songs are a good example of sound communication being used for territorial defense and species recognition. A territory may be defined as the area defended by an animal. This area centers around the animal's breeding ground, nest site, and sources of food or other needs. The use of sound to defend a territory is exemplified in the following experiment. If silent models of wood thrushes are set up within the territory of a given thrush, they would be attacked by that thrush. If, however, a loudspeaker is set up next to the models, and the characteristic species song of another species played, the thrush will not attack. This is because males attack only other males of their own species, since they are the most threatening to their territories.

Singing also plays a role in the establishment of territories. A male bird chooses an area and sings loudly in order to warn away other males. During the spring, when boundaries are being established, the male thrushes that sing most loudly and forcefully are those who successfully acquire the largest territories.

9.6   Receptors in Other Organisms

A) **Protozoans** – The plasma membrane in these organisms is the receptor for external stimuli.

B) **Hydra** – The hydra possesses nerve endings in the ectoderm and endoderm layers that are sensitive to light, touch, and chemicals.

C) **Grasshopper** – The grasshopper possesses many sense organs which are sensitive to external stimuli. The compound and single eyes are sensitive to bright light and dim light, respectively. Hairs which are found in various areas on the body are sensitive to touch. The tympanic membranes, located on the grasshopper's abdomen, transmit sound vibrations to receptor cells.

D) **Earthworm** – The nerve endings in the skin of the earthworm possess receptors for touch, chemicals, and temperature changes. Light receptors are mainly found in the head and tail areas.

CHAPTER 10

The Nervous System

10.1 The Nervous System

The nervous system is a system of conduction that transmits information from receptors to appropriate structures for action.

10.2 Neurons

The neuron is the unit of structure that conducts electrochemical impulses over a certain distance. In many neurons, the nerve impulses are generated in the dendrites. These impulses are then conducted along the axon, which is a long fiber. A phospholipid myelin sheath covers the axon, and serves to increase impulse transmission.

A) **Sensory Neurons** – Sensory neurons conduct impulses from receptors to the central nervous system.

B) **Interneurons** – Interneurons of the central nervous system are primarily found in the brain and spinal cord. They form the intermediate link in the nervous system pathway.

C) **Motor Neurons** – Motor neurons conduct impulses from the central nervous system to the effectors which are muscles and glands. They will bring about the responses to the stimulus.

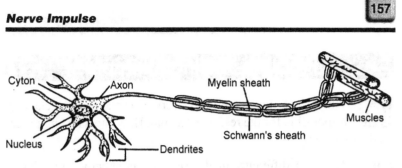

Figure 10.1 The structure of a neuron

Problem Solving Example:

How are many kinds of neurons are there? What functions are performed by each type?

Neurons are classified on the basis of their functions into sensory, motor, or interneurons. Sensory neurons, or afferent neurons, are either receptors that receive impulses or connectors of receptors that conduct information to the central nervous system (brain and spinal cord). Motor neurons, or efferent neurons, conduct information away from the central nervous system to the effectors, for example, muscles and glands. Interneurons, which connect two or more neurons, usually lie wholly within the central nervous system. Therefore, they have both axonal and dendritic ends in the central nervous system. In contrast, the sensory and motor neurons generally have one of their endings in the central nervous system and the other close to the periphery of the body.

10.3 Nerve Impulse

The nerve impulse is the signal that is transmitted from one neuron to another. When a neuron is not stimulated, the outside of the neuron is positively charged and the inside is negatively charged. However, when there is neuron stimulation, the inside of the neuron is temporarily positively charged and the outside is temporarily negatively charged. This marks the beginning of the generation and flow of the nerve impulse.

Problem Solving Example:

 What is a resting potential? Describe the chemical mechanism responsible for the resting potential. How can a resting potential be detected?

There is a difference in electrical potential between the inside and the outside of all living cells. For example, the potential difference across the plasma membrane of the neuron is measured to be about 60 millivolts, the inside being negative with respect to the outside. This potential difference is called the resting potential.

The chemical basis for the resting potential is as follows. By active transport, an energy-requiring process that transfers substances across the plasma membrane against their concentration gradients, the concentration of potassium (K^+) ions is kept higher inside the cell than outside. At the same time, there is a lower concentration of sodium (Na^+) ions in the cell interior than the exterior. Moreover, in the resting state the permeability of the plasma membrane is different to K^+ and Na^+ ions. The membrane is more permeable to K^+ ions than to Na^+ ions. Hence, K^+ ions can move across the membrane by simple diffusion to the outside more easily than Na^+ ions can move in to replace them. Because more positive charges (K^+) leave the inside of the cell than are replenished (by Na^+), there is a net negative charge on the inside and a net positive charge on the outside. An electrical potential is established across the membrane. This potential is the resting membrane potential.

We can measure the resting potential by placing one electrode, insulated except at the tip, inside the cell and a second electrode on the outside surface and connecting the two with a sensitive galvanometer. The reading on the galvanometer should be approximately 60 millivolts if the cell tested is a neuron. (Different types of cells, such as skeletal and cardiac muscle cells, vary in their values of resting potential.) Note, however, that if both electrodes are placed on the outside surface of the cell, no potential difference between them is registered because all points on the outside are at equal potential. The same is true if both electrodes are placed on the inside surface of the plasma membrane.

10.4 Synapse

The synapse is the junction between the axon of one neuron and the dendrite of the next neuron. An impulse is transmitted across the synaptic gap by a specific chemical neurotransmitter such as acetylcholine. When a nerve impulse reaches the end brush of the first axon, the end brush secretes neurotransmitter acetylcholine into the synaptic gap. It is here that the neurotransmitter acetylcholine changes the permeability of the dendrite's membrane of the second neuron. As soon as the neurotransmitter acetylcholine is no longer needed, it is decomposed by an enzyme neurotransmitter.

Problem Solving Example:

 What are the chemical and physical processes involved in transmission at the synapse?

The nervous system is composed of discrete units, the neurons, yet it behaves like a continuous system of transmission of impulses. For this to occur, there have to be functional connections between neurons. These connections are known as synapses. A synapse is an anatomically specialized junction between two neurons, lying adjacent to each other, where the electric activity in one neuron (the presynaptic neuron) influences the excitability of the second (the postsynaptic neuron). At the synapse, the electric impulse is transformed into a chemical form of transmission.

Chemical transmission at the synapse involves the processes of neurosecretion and chemoreception. The arrival of a nerve impulse at the axon terminal stimulates the release of a specific chemical substance, which has been synthesized in the cell body and stored in the tip of the axon, into the narrow synaptic space between the adjacent neurons. This process constitutes neurosecretion. The chemical secreted, known as a neurotransmitter, can cause local depolarization of the membrane of the postsynaptic region and thus can transmit the excitation to the adjacent neuron. Chemoreception is the process in which the neurotransmitter becomes attached to specific molecular sites on the membrane of the dendrite (post synaptic region), producing a change in the properties of the cell membrane so that a new impulse is established.

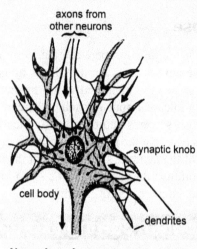

Nerve impulse across a synapse

The neurotransmitter passes from the presynaptic axon to the postsynaptic dendrite by a simple diffusion across the narrow space, called the synaptic cleft, separating the two neurons involved in the synapse. The synaptic clefts have been measured under the electron microscope to be about 200 Å in width. Diffusion is rapid enough to account for the speed of transmission observed at the synapse. After the neurotransmitter has exerted its effect on the postsynaptic membrane, it is promptly destroyed by an enzyme called acetylcholinesterase. This destruction is of critical importance. If the acetylcholine were not destroyed, it would continue its stimulatory action indefinitely and all control would be lost.

10.5 Reflex Arc

The reflex arc is the unit of function of the nervous system. It is formed by a sequence of sensory neurons, interneurons, and motor neurons which conduct the nerve impulses for the given reflex.

Problem Solving Example:

Describe and give an example of a reflex arc.

To understand what a reflex arc is, we must know something about reflexes. A reflex is an innate, stereotyped, automatic response to a given stimulus. A popular example of a reflex is the knee jerk. No matter how many times we rap on the tendon of a person's knee cap, his leg will invariably straighten out. This experiment demonstrates one of the chief characteristics of a reflex: fidelity of repetition.

Reflexes are important because responses to certain stimuli have to be made instantaneously. For example, when we step on something sharp or come into contact with something hot, we do not wait until the pain is experienced by the brain and then after deliberation decide what to do. Our responses are immediate and automatic. The part of the body involved is being withdrawn by reflex action before the sensation of pain is experienced.

A reflex arc is the neural pathway that conducts the nerve impulses for a given reflex. It consists of a sensory neuron with a receptor to detect the stimulus, connected by a synapse to a motor neuron, which is attached to a muscle or some other tissue that brings about the appropriate response. Thus, the simplest type of reflex arc is termed monosynaptic because there is only one synapse between the sensory and motor neurons. Most reflex arcs include one or more interneurons between the sensory and motor neurons.

An example of a monosynaptic reflex arc is the knee jerk. When the tendon of the knee cap is tapped, and thereby stretched, receptors in the tendon are stimulated. An impulse travels along the sensory neuron to the spinal cord where it synapses directly with a motor neuron. This latter neuron transmits an impulse to the effector muscle in the leg, causing it to contract, resulting in a sudden straightening of the leg.

10.6 The Human Nervous System

A) **The Central Nervous System** – The brain, spinal cord, and retina constitute the central nervous system. The brain is divided into three regions: the forebrain, the midbrain, and the hindbrain. Each of these regions has a specific function attributed to the particular lobe.

1) Brain

a) **Forebrain** – The cerebrum is the most prominent part of the forebrain and is divided into two hemispheres. It also has four major areas known as the sensory area, motor area, speech area, and association area. The thalamus, hypothalamus, pineal gland, and part of the pituitary gland are also part of the human forebrain. (See figure on following page.)

b) **Midbrain** – The midbrain is one of the smallest regions of the human brain. Its main function is to relay nerve impulses to other brain regions. It also aids in the maintenance of balance.

c) **Hindbrain** – The medulla oblongata, pons and the cerebellum are the two main regions of the hindbrain.

1) **Medulla Oblongata** – controls reflex centers for respiration and heartbeat, coughing, swallowing, and sneezing.

2) **Cerebellum** – coordinates locomotor activity in the body initiated by impulses originating in the forebrain.

2) **Spinal Cord** – The spinal cord runs from the medulla down through the backbone. Throughout its length, it is enclosed by three meninges and by the spinal column vertebrae. Running vertically in the spinal cord center is a narrow canal filled with cerebrospinal fluid.

The spinal cord contains control centers for reflex acts below the neck, and it provides the major pathway for impulses be-

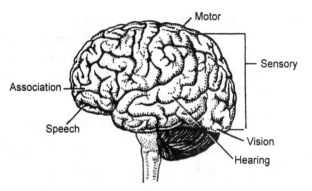

Figure 10.2 The major areas of the cerebrum

tween the peripheral nervous system and the brain. It is also a connecting center between sensory and motor neurons.

B) **The Peripheral Nervous System** – The peripheral nervous system is composed of nerve fibers which connect the brain and the spinal cord (central nervous system) to the sense organs, glands, and muscles. It can be subdivided into the somatic nervous system and the autonomic nervous system.

   1) **Somatic Nervous System** – The somatic nervous system consists of nerves which transmit impulses from receptors to the central nervous system and from the central nervous system to the skeletal muscles of the body.

   2) **Autonomic Nervous System** – The autonomic nervous system is composed of sensory and motor neurons which run between the central nervous system and various internal organs such as the heart, glands, and intestines. It regulates internal responses which keep the internal environment constant. It is subdivided into two smaller systems.

      a) **Sympathetic System** – A branch of the autonomic nervous system with motor neurons arising from the spinal cord. This system accelerates heart beat rate, constricts arteries, slows peristalsis, relaxes the bladder, dilates breathing passages, dilates the pupil, and increases certain secretion.

b) **Parasympathetic System** – A branch of the autonomic nervous system which consists of fibers arising from the brain. The effectors on the organs innervated by the parasympathetic system are opposite to the effects of the sympathetic system.

Table 10.1 Actions of the autonomic nervous system

Organ Innervated	Sympathetic Action	Parasympathetic Action
Heart	Accelerates heartbeat	Slows heartbeat
Arteries	Constricts arteries	Dilates arteries
Lungs	Dilates bronchial passages	Constricts bronchial passage
Digestive Tract	Slows peristalsis rate	Increases peristalsis rate
Eye	Dilates pupil	Constricts pupil
Urinary Bladder	Relaxes bladder	Constricts bladder

10.7 The Nervous System of Other Organisms

A) **Protozoans** – Protozoans have no nervous system; however, their protoplasm does receive and respond to certain stimuli.

B) **Hydra** – The hydra possesses a simple nervous system which has no central control, known as a nerve net. A stimulus applied to a specific part of the body will generate an impulse which will travel to all body parts.

C) **Earthworm** – The earthworm possesses a central nervous system which includes a brain, a nerve cord, which is a chain of ganglions, sense organs, nerve fibers, muscles, and glands.

D) **Grasshopper** – The grasshopper's nervous system consists of ganglia bundled together to form the peripheral nervous system. The ganglia of the grasshopper are better developed than in the earthworm.

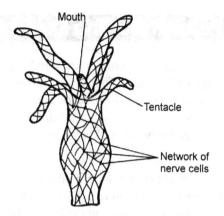

Figure 10.3 Nerve net of the hydra

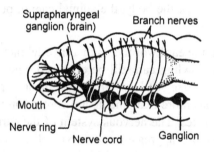

Figure 10.4 The earthworm's nervous system

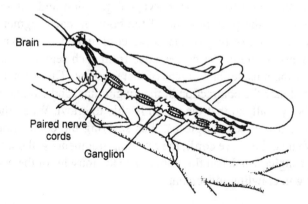

Figure 10.5 The nervous system of the grasshopper

Problem Solving Example:

 The earthworm has a central nerve cord running along the entire length of the body. When an earthworm is cut into several parts, severing its nerve cord in the process, each part will go on crawling around for some time. How can each fragment carry out crawling movement even after the nerve cord is severed?

A The arrangement of the nervous system in lower invertebrates is quite different from that in higher vertebrates like humans. In higher vertebrates, there is an expanded and highly developed anterior end of the spinal cord, forming the brain. The brain coordinates and regulates the activities of the entire body. When the brain is separated from the rest of the body, the animal cannot perform any complicated functions.

However, in the earthworm, a higher invertebrate, coordinated activities such as crawling can still be observed when the body is cut into several transverse sections. This movement is possible because the earthworm has more than one neural center controlling and coordinating its activities. The nervous system of the earthworm consists of a large, two-lobed aggregation of nerve cells, called the brain, located just above the pharynx in the third segment, and a subpharyngeal ganglion just below the pharynx in the fourth segment. A nerve cord connects the brain to the subpharyngeal ganglion and extends from the anterior to the posterior end of the body. In each segment of the body there is a swelling of the nerve cord, called a segmental ganglion. Sensory and motor nerves arise from each segmental ganglion to supply the muscles and organs of that segment. The segmental ganglia coordinate the contraction of the longitudinal and circular muscles of the body wall, so that the worm is able to crawl. When the earthworm is cut into several pieces, thus severing the connection to the brain (that is, the nerve cord), the resulting fragments still contain segmental ganglia which can fire impulses to the muscles of the body wall, resulting in crawling movement.

Quiz: Stimulus Receptors–
The Nervous System

1.   Proprioception involves

    (A)   coordination of the eyes and ears.

    (B)   the ability to repeat something that is spoken.

    (C)   the awareness of body position involving muscles, tendons, and the cerebellum.

    (D)   the ability to develop short and long-term memory.

2.   Which of the following is incorrect?

    (A)   Myopia (nearsightedness) is a condition caused by light rays converging in front of the retina, and can be corrected by using a concave lens in front of the eye.

    (B)   Hypermetropia (farsightedness) is a condition caused by light rays converging behind the retina, and can be corrected by use of a convex lens in front of the eye.

    (C)   Astigmatism is a condition caused by an irregularly shaped cornea.

    (D)   Myopia and hypermetropia are both caused by a combination of irregular focusing properties and an irregularly shaped cornea.

3.   The fovea consists of

    (A)   equal amounts of rods and cones.

    (B)   more rods than cones.

    (C)   more cones than rods.

    (D)   only cones.

4. Which of the following statements is incorrect?

   (A) Fibers from the parasympathetic division of the autonomic nervous system (ANS) arise from the brainstem
   and sacral portions of the spinal cord, and, as an example of one of its actions, decelerate the heart rate.

   (B) Fibers from the sympathetic division of the ANS arise
   from the thoracic and lumbar regions of the spinal
   cord, and, as an example of one of its actions, accelerate the heart rate.

   (C) Both the parasympathetic and sympathetic divisions
   of the ANS arise from all five regions of the spinal
   cord, and act cooperatively to accelerate the heart rate.

   (D) The parasympathetic division of the ANS is responsible for the so-called "rest and digest" reflex, while
   the sympathetic division of the ANS is responsible for
   the so-called "fight or flight" reflex.

5. Which of the following statements is *completely* correct?

   (A) Actions of the parasympathetic division of the ANS
   include slowing of the heart rate, dilation of blood vessels, dilation of the pupils, and bladder constriction.

   (B) Actions of the sympathetic division of the ANS include increased non-digestive glandular secretion, dilation of the pupils, decreased digestive gland secretions, and dilation of airway passages.

   (C) Actions of the sympathetic division of the ANS include constriction of blood vessels, inhibition of peristalsis for digestion, constriction of the bladder, and
   acceleration of the heart rate.

   (D) Actions of the parasympathetic division of the ANS
   include decreased non-digestive glandular secretion,
   constriction of airway passages, constriction of the
   pupils, and acceleration of peristalsis for digestion.

6. Which of the following is the correct explanation of the composition and function of the myelin sheath?

    (A) The myelin sheath is a secretion product of Schwann cells composed of protein. Its function is that of synthesis of neurotransmitter substance.

    (B) The myelin sheath is a secretion product of Schwann cells composed of carbohydrates. It functions to slow the speed of nerve conduction.

    (C) The myelin sheath is the coiled membrane of a Schwann cell, and is mainly composed of phospholipids. It functions to increase the speed with which a nerve impulse is transmitted.

    (D) The myelin sheath is composed of connective tissue and has no relationship to Schwann cells. Its function is to act as a protective covering for nerves.

7. Which of the following is *not* involved in the process of synaptic transmission?

    (A) The release of a neurotransmitter from synaptic vesicles at the pre-synaptic neuron

    (B) The destruction of the post-synaptic membrane after the neurotransmitter has come in contact with it

    (C) Diffusion of the neurotransmitter across the synaptic cleft

    (D) Destruction of the neurotransmitter after transmission of the impulse has taken place

8. Which of the following statements is *incorrect*?

    (A) A peripheral nerve, even though composed of many nerve fibers, will elicit an all-or-none response with only one threshold value.

    (B) The all-or-none response means that a nerve fiber will

"fire" only if the threshold value of stimulation is reached.

(C) Once threshold value is reached, the fiber will fire to the same level regardless of how strong the stimulus is. The frequency of firing will change though, as the stimulus gets stronger.

(D) As a stimulation gets more intense, more individual nerve fibers in a peripheral nerve will fire, exhibiting their own all-or-none responses.

9. Neurons that conduct signals away from the central nervous system are classified as

(A) afferent.

(B) associative.

(C) internuncial.

(D) motor.

10. When a neuron is not stimulated

(A) the outside of the neuron is negatively charged and the inside is positively charged.

(B) the outside of the neuron is positively charged and the inside is negatively charged.

(C) both inside and outside of the neuron are negatively charged.

(D) both inside and outside of the neuron are positively charged.

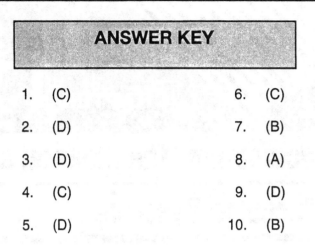

ANSWER KEY

1.	(C)	6.	(C)
2.	(D)	7.	(B)
3.	(D)	8.	(A)
4.	(C)	9.	(D)
5.	(D)	10.	(B)

CHAPTER 11

Locomotion: The Skeletal and Muscular Systems

11.1 The Skeletal System in Humans

The cartilage, ligaments, and a skeleton composed of 206 bones make up the skeletal system of humans.

A) **Cartilage** – Cartilage is a soft material present at the ends of bones, especially at joints.

B) **Ligaments** – Ligaments are strong bands of connective tissue which bind one bone to another bone.

C) **Skeleton** – The bones of the skeleton serve five important functions:

  1) Allowing movements of parts of the body.

  2) Supporting various organs of the body.

  3) Supplying the body with red blood cells and some white blood cells.

4) Protecting internal organs.

5) Storing calcium and phosphate salts.

The skeleton is composed of the vertebral column, the skull, the limbs, the breastbone, and the ribs.

Problem Solving Example:

Besides their function in locomotion and support, bones also serve several other important functions. What are they?

Bones are an important reservoir for certain minerals. The mineral content of bones is constantly being renewed. Roughly all the mineral content of bone is removed and replaced every nine months. Calcium and phosphorus are especially abundant in the bones and must be maintained in the blood at a constant level. When the diet is low in these minerals, they are withdrawn from the bones to maintain the proper concentration in the blood. In the absence of physical stress these minerals pass from the bones into the blood. This elevates the blood concentration of these minerals to a very high level, which may ultimately lead to the development of kidney stones. Before special stress exercise programs were developed, astronauts in space often became victims of this type of kidney trouble.

During pregnancy, when the demand for minerals to form bones of a growing fetus is great, a woman's own bones may become depleted unless her diet contains more of these minerals than is normally needed. During starvation, the blood can draw on the storehouse of minerals in the bones and maintain life much longer than would be possible without this means of storage.

Bones are also important in that they give rise to the fundamental elements of the circulatory system. Bone marrow is the site of production of lymphocyte precursor cells, which plays an integral role in the body's immune response system. Red blood cells, or erythrocytes, also originate in the bone marrow. As the erythrocytes mature, they accumulate hemoglobin, the oxygen carrier of blood. Mature erythrocytes, however, are incomplete cells lacking nuclei and the metabolic machinery to synthesize new protein. They are released into the bloodstream, where they circulate for approximately 120 days before being destroyed by the phagocytes. Thus, the bone marrow must perform the constant task of maintaining the level of erythrocytes for the packaging of hemoglobin.

11.2 The Muscular System in Humans

A) Kinds of Muscles

1) **Smooth Muscle** – Smooth muscle is found in the walls of the hollow organs of the body.

2) **Cardiac Muscle** – Cardiac muscle is the muscle that comprises the walls of the heart.

3) **Skeletal Muscle** – Skeletal muscles are muscles attached to the skeleton. They are also known as striated muscles.

B) **The Structure of Muscles and Bones** – Bones move only when there is a pull on the muscles attached to the bone. A single skeletal muscle consists of:

1) **Tendon** – A tendon is a band of strong, connective tissue which attaches muscle to bone.

2) **Origin** – The origin is one end of the muscle which is attached to a bone that does not move when the muscle contracts.

3) **Insertion** – The insertion is the other end of the muscle which is attached to a bone that moves when the muscle contracts.

4) **Belly** – The belly is the thickened part of the muscle which contracts and pulls.

C) **Skeletal Muscle Activation** – The nervous system controls skeletal muscle contraction. End brushes of motor neurons come in contact with muscle fibers at the motor end plate, a synapse-like junction. Muscle contraction occurs when acetylcholine is discharged on the muscle fiber surface after the impulse reaches the motor end plate.

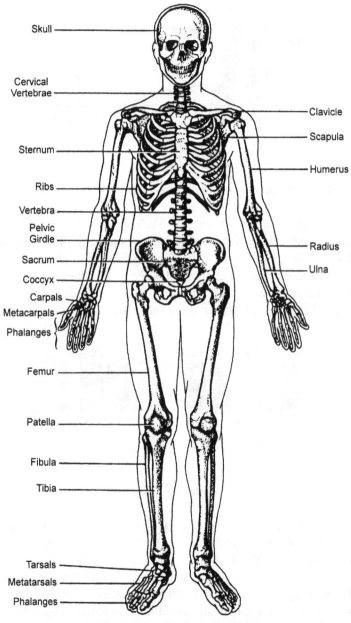

Figure 11.1 The human skeleton

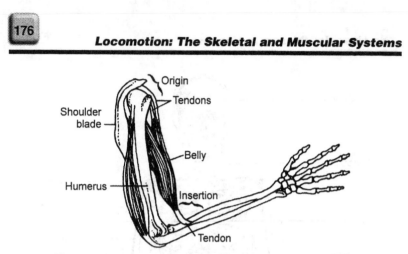

Fig 11.2 The mechanism of movement of the upper arm

D) **Structure of a Muscle Fiber** – Skeletal muscle is composed of long fibers whose cytoplasm possesses alternating light and dark bands. These bands are part of fibrils which lie parallel to one another. The dark bands are termed A-bands and the light bands are the I-bands. The H-band bisects the A-bands while the Z-line bisects the I-band. Skeletal muscles are often multi-nucleated.

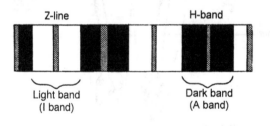

Figure 11.3 Single skeletal muscle fibril

E) **Chemical Composition of Muscle Contraction** – Thick filaments that make up the A-band are composed of the protein myosin. The thin filaments extend in either direction from the Z-line and are composed of the protein actin. When an impulse enters a muscle fiber, energy is released from ATP molecules. A complex combination of actin and myosin, called actinomyosin, is then formed. The fiber contracts with actinomyosin formation. ATP is needed for muscle contraction. The mitochondria, which are present in muscle cells, release the energy needed to form ATP.

Problem Solving Example:

Q What is meant by the term muscle fatigue?

A A muscle that has contracted several times, exhausting its stored supply of organic phosphates and glycogen, will accumulate lactic acid. It is unable to contract any longer and is said to be "fatigued." Fatigue is primarily induced by this accumulation of lactic acid, which correlates closely with the depletion of the muscle stores of glycogen. Fatigue, however, may actually be felt by the individual before the muscle reaches the exhausted condition.

The spot most susceptible to fatigue can be demonstrated experimentally. A muscle and its attached nerve can be dissected out and the nerve stimulated repeatedly by electric shock until the muscle no longer contracts. If the muscle is then stimulated directly by placing electrodes on the muscle tissue, it will contract. With the proper device for detecting the passage of nerve impulses, it can be shown that upon fatigue, the nerve leading to the muscle is not fatigued, but remains capable of conduction. Thus, since the nerve is still conducting impulses and the muscle is still capable of contracting, the point of fatigue must be at the junction between the nerve and the muscle, where nerve impulses initiate muscle contraction. Fatigue is then due in part to an accumulation of lactic acid, in part to depletion of stored energy reserves, and in part to breakdown in neuromuscular junction transmission.

11.3 Locomotion in Lower Organisms

A) **Amoeba** – Locomotion in the amoeba is carried out by pseudopods, the primary adaptation for locomotion.

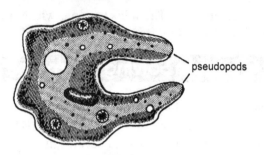

Figure 11.4 Pseudopods of the amoeba

B) **Paramecium** – The wave-like beating of cilia provide the paramecium with an adaptation for locomotion.

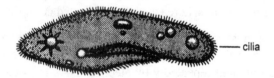

Figure 11.5 Cilia surrounding paramecium

C) **Euglena** – The euglena moves by means of a flagellum and by contraction and elongation of its body.

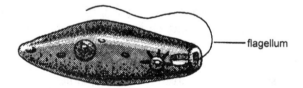

Figure 11.6 Flagellum of the euglena

Problem Solving Example:

Describe locomotion in the amoeba.

The locomotion of the amoeba is considered to be the simplest type of animal locomotion. A moving amoeba sends out a projection, termed a pseudopodium. Following this, the organism advances as the inner, granular, gel-like endoplasm flows into the pseudopodium. Two or three pseudopodia may be formed simultaneously, but ultimately one will become dominant for a time. As new pseudopodia are formed, the old ones withdraw into the general body region. In its locomotion, the amoeba often changes its course in response to environmental stimulation by forming a new dominant pseudopodium on the opposite side, thus moving in a very irregular fashion.

Currently there is no fully complete explanation for the physical and chemical changes which are involved in amoeboid movement. The theory accepted by zoologists at the present time is based on changes in the texture of the cytoplasm. As a result of some initial stimulus, ectoplasm, the outer clear, thin layer of the organism, undergoes a liquefaction and becomes endoplasm, which is gel-like. As a result of this change, internal pressure builds up and causes the endoplasm to flow out at this point, forming a pseudopodium. In the interior of the pseudopodium, the endoplasm flows forward along the line of progression; around the periphery, endoplasm is converted to ectoplasm, thereby building up and extending the sides of the pseudopodium like a sleeve. At the posterior of the body, ectoplasm is assumed to be undergoing conversion to endoplasm. During this entire process, energy consumption is known to have taken place.

Quiz: Locomotion–The Skeletal and Muscular Systems

1. Transverse light and dark bands form a regular pattern along

    (A) all muscle fibers.

    (B) only smooth muscle fibers.

    (C) skeletal and cardiac muscle fibers.

    (D) skeletal and smooth muscle fibers.

2. The A-band is

    (A) the thick dark band of myofilaments.

    (B) composed of the protein actin.

    (C) the thin myofilament containing the protein myosin.

    (D) None of the above

3. Muscle contraction occurs when

    (A) acetylcholine is released by muscle fibers.

    (B) the Z-line is crossed.

    (C) actin and myosin combine.

    (D) acetylchloline is degraded.

4. Muscle fatigue occurs when

    (A) nerve fibers become exhausted from repeated firing.

    (B) lactic acid builds up so much that muscle fibers can no longer contract.

(C) lactic acid is so depleted that muscle fibers can no longer contract.

(D) nerve fibers run out of their supply of acetylchloline.

5. Choose the correct statement:

(A) Skeletal muscle fibers are multinucleated.

(B) Smooth muscle shows a striated banding pattern.

(C) Cardiac muscle is striated and has a single nucleus.

(D) All of the above are correct.

6. Which of the following supplies energy for muscular contraction?

(A) ADP

(B) ATP

(C) lactic acid

(D) myosin

7. Which of the following is not a function of bone?

(A) replacement of certain minerals

(B) production of lymphocyte precursor cells

(C) production of erythrocytes

(D) All of the above are functions of bone.

8. Which of the following best explains the probable mechanism of amoeboid movement?

(A) The simultaneous conversion of both ectoplasm and endoplasm to a gel-like mass causes the formation of a pseudopodia. This acts as a foot-like device, as the amoeba uses it to move around.

(B) The simultaneous conversion of both ectoplasm and endoplasm to a liquid-like mass causes pressure to build up on the rim of the amoeba, forming a pseudopodia.

(C) Neither ectoplasm nor endoplasm changes its original consistency. A pseudopodia is formed due to a change in internal pressure inside the interior of the amoeba. The pseudopodia is then used by the amoeba to move around.

(D) Ectoplasm moves from outside to inside along the periphery of the cell, causing pressure to build up inside the amoeba. This causes the endoplasm to move out to the periphery, causing the formation of a pseudopodium. The endoplasm then continues flowing outward, until its flow is fountain-like, moving back toward the cell interior. The process then begins again.

9. Which of the following best describes ciliary movement in a Paramecium?

(A) Simple back and forth movement

(B) Movement similar to a rowing motion

(C) A highly coordinated motion consisting of a power stroke and a recovery stroke

(D) Movement that begins at the tip of the cilia and moves downward in a wave-like motion

10. Which tissue has the ability to repair itself most rapidly?

(A) Epithelium

(B) Connective

(C) Muscle

(D) Bone

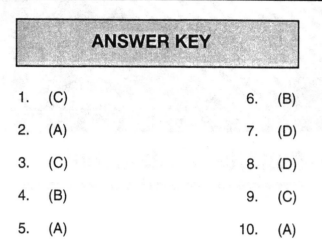

ANSWER KEY

1.	(C)	6.	(B)
2.	(A)	7.	(D)
3.	(C)	8.	(D)
4.	(B)	9.	(C)
5.	(A)	10.	(A)

CHAPTER 12

Elements of Behavior

12.1 Learned Behavior

Conditioning, habits, and imprinting are all specific types of behavior that are learned and acquired as the result of individual experiences.

A) **Conditioning** – Conditioned behavior is a response caused by a stimulus different from that which originally triggered the response. Experiments conducted by Pavlov on dogs demonstrate conditioning of behavior.

---

ringing of bell (stimulus 1) $\rightarrow$ barking (response 1)

food + ringing of bell (stimulus 2) $\rightarrow$ saliva flow (response 2)

ringing of bell (stimulus 1) $\rightarrow$ saliva flow (response 2)

Figure 12.1 Conditioned behavior in Pavlov's dogs

B) **Habits** – Habit behavior is learned behavior that becomes automatic and voluntary as a consequence of repetition. When an action is constantly repeated, the amount of thinking is reduced because impulses pass through the nerve pathways more quickly. The behavior soon becomes automatic.

C) **Imprinting** – Imprinting involves the establishment of a fixed pathway in the nervous system by the stimulus of the very first object that is seen, heard, or smelled by the particular organism. The research of Konrad Lorenz with newly hatched geese demonstrates this type of learning.

Problem Solving Example:

Using conditioning, how can one determine the sensitivity of a dog's hearing to sounds of different frequencies?

Dogs can usually discriminate between sounds of different frequencies or pitches (measured in cycles per second, cps). To determine how well they discriminate, one can use operant conditioning. By rewarding or punishing a particular act when it occurs, the probability that it will occur again is either increased or decreased. For example, one can expose the dog to a sound frequency of 1,000 cps. The dog is then rewarded whenever it responds to a sound of that frequency by barking. Soon the dog barks whenever it hears a sound of 1,000 cps. In the same manner, the dog can be conditioned to raise its paw whenever it hears sounds of another frequency. By making this frequency closer and closer to 1,000 cps, and observing the dog's response, either a bark or a paw lift, one can determine the smallest differences in frequency between which the dog can discriminate. A point is eventually reached beyond which the dog is unable to discriminate between a "bark" pitch and a "lift-paw" pitch. Dogs have been found to be capable of discriminating between two sounds that differ by only two cycles per second.

Generalization is the response that is contradictory to discrimination. This occurs when an animal conditioned to one stimulus responds in the same manner to a similar, though different stimulus. For example, a bird conditioned to peck at a dark blue spot may peck at a violet spot when the blue one is unavailable.

12.2 Innate Behavior

Taxis, reflexes, and instincts are all specific types of behavior that are inborn and involuntary.

A) **Taxis** – Taxis is the response to a stimulus by automatically moving either toward or away from the stimulus.

   1) **Phototaxis** – Photosynthetic microorganisms move toward light of moderate intensity.

   2) **Chemotaxis** – Organisms move in response to some chemical.

Figure 12.2 E. coli bacteria congregate near a specific chemical

B) **Reflex** – A reflex is an automatic response to a stimulus in which only a part of the body is involved; it is the simplest inborn response.

   The knee jerk is a stretch reflex that is a response to a tap on the tendon below the knee cap which stretches the attached muscle. This tapping activates stretch receptors. Stretching a spindle fiber triggers nerve impulses.

C) **Instinct** – An instinct is a complex behavior pattern which is un-learned and automatic and is often beneficial in adapting the individual to his or her environment.

Nest-making of birds and web-spinning by spiders are examples of instinctive behavior.

1) **Instinct of Self-preservation** – This is characterized by "fight or flight" behavior of animals.

2) **Instinct of Species-preservation** – This is characterized by the instinctive behavior of the animal not to escape or fight, but to find a safer area for habitation.

3) **Releasers** – The releasers are signals which possess the ability to trigger instinctive acts.

Problem Solving Example:

Q Black-headed gulls remove broken eggshells from their nests immediately after the young have hatched. Explain the biological significance of this behavior.

A Behavior that contributes to the survival and reproduction of the animal is adaptive. Since those animals which demonstrate adaptive behavior survive longer and successfully raise more offspring, it is more likely that the behavior patterns of these animals will be continued throughout further generations.

The eggshell-removing habit of black-headed gulls is one such adaptive behavior. The gulls do not remove only broken shells, but also any conspicuous object placed in the nest during the breeding season, as a defense mechanism against visual predators. When investigators placed conspicuous objects along with eggs in nests, these nests were robbed of eggs (by other gulls) more often than nests having only eggs. Thus eggshell-removing behavior is significant in that it reduces the chances of a nest being robbed, thus enhancing the survival of offspring. The gulls that evoke this behavioral pattern will successfully raise more offspring than those who do not, and thus this behavioral pattern will tend to be passed on.

12.3 Voluntary Behavior

Voluntary behavior includes activities under direct control of the will such as learning and memory.

A) **Learning** – Intelligence measures the ability to learn and properly establish new patterns of behavior. Humans demonstrate the highest degree of intelligence among all animals. This is due, in part, to the highly developed cerebrum which contains a great quantity of nerve pathways and neurons.

B) **Memory** – All learning is dependent upon one's memory. Memory is essential for all previous learning to be retained and used.

Problem Solving Example:

What is behavior? Are behavior patterns solely inherited or can they be modified?

The term behavior refers to the patterns by which organisms survive and reproduce. The more an organism must actively search the environment to maintain life, the more advanced are its behavior patterns. Thus a deer shows more complex behavior patterns than does a planaria. Behavior is not limited to the animal kingdom; plants display simple behavior, such as the capture of prey by the Venus flytrap.

Behavior has a genetic or hereditary basis that is controlled by DNA. The anatomical aspects of both the nervous and endocrine systems are chiefly determined by the genetic composition. These two systems are responsible for most behavioral phenomena. In general, an organism's behavior is principally an expression of the capabilities of its nervous system, modified to various extents by its endocrine system.

Behavioral development is often influenced by experience. This modification of behavioral patterns through an organism's particular experiences is called learning. For example, some birds transmit their species-specific song to their offspring solely by heredity, while others must experience the singing of the song by other members of their species in order for them to acquire the song.

Both inheritance and learning are therefore fundamental in the determination of behavioral patterns. Inheritance determines the limits within which a pattern can be modified. The nervous system of a primitive animal can limit the modification of existing behavior patterns, allowing only simple rigid behavioral responses. In highly developed animals such as humans, there are fewer purely inherited limits, permitting learning to play a significant role in determining behavior.

Quiz: Elements of Behavior

1. Which of the following is capable of performing a reflex behavior?

    (A) any eukaryote.

    (B) any multicelled eukaryote.

    (C) all organisms except viruses.

    (D) any organisms with a nervous system.

2. A particular act is rewarded (or punished) when it occurs, therefore increasing (or decreasing) the probability of the act being repeated. This is an example of

    (A) habituation.

    (B) operant conditioning.

    (C) a conditioned reflex.

    (D) classic reflex conditioning.

3. A wood louse becomes restless after conditions become dry. It moves randomly until it finds a moist environment. This is

    (A) an example of chemotaxis.

    (B) not an example of taxis because the louse moved in a random direction.

    (C) an example of taxis because the louse moved in a random direction.

    (D) not an example of phototaxis because the louse moved toward a chemical.

4. Which of the following identifies a reflex behavior?

   (A) a mother bird bringing food to its chicks.

   (B) a dog that barks at strangers.

   (C) blinking your eyes when you hear an unexpected loud noise.

   (D) a dog that learns to roll over.

5. Which of the following would be the most convincing proof that learning was taking place?

   (A) a bird that sings a new song that it has never heard before

   (B) a bird that sings a new song only after hearing it from another bird

   (C) a bird that sings only after being exposed to the singing of other birds

   (D) a bird that sings without being exposed to the singing of other birds

6. When an animal learns to make a strong association with another organism during a brief period early in development, this is called

   (A) conditioning.

   (B) imprinting.

   (C) reinforcement.

   (D) habituation.

Use the following choices to answer Questions 7-10

    (A) conditioning

    (B) habituation

    (C) imprinting

    (D) insight

7. Turning response to a stimulus such as light or gravity, particularly by different growth patterns in plants

8. The association through reinforcement of a response to a stimulus with which it was not previously associated

9. The strong attachment of newly-born animals with the first moving object they observe

10. A simple learning type with a gradual decline in response to a stimulus through repeated exposure

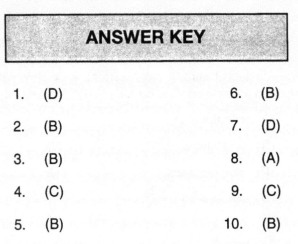

ANSWER KEY

1.	(D)		6.	(B)
2.	(B)		7.	(D)
3.	(B)		8.	(A)
4.	(C)		9.	(C)
5.	(B)		10.	(B)

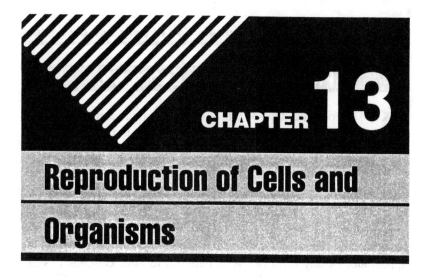

CHAPTER 13

Reproduction of Cells and Organisms

13.1 The Origin of Life

A) **Spontaneous Generation** – Discredited idea which proposed that new generations of living organisms arose from non-living matter.

B) **Francesco Redi** – Italian physician who discredited the idea of spontaneous generation with his own experiments from which he concluded that flies arise from eggs that are laid by parent flies on decaying meat, and not from the meat.

C) **Lazzaro Spallanzini** – Italian priest who also helped to discredit the idea of spontaneous generation by conducting experiments which concluded that bacteria arise from other bacteria and not from non-living vegetable juice.

D) **Louis Pasteur** – French chemist who convincingly discredited the idea of spontaneous generation by proving that bacteria arise from other parent bacteria.

Problem Solving Example:

Q   Describe the steps by which simple inorganic substances may have undergone chemical evolution to yield the complex system of organic chemicals we recognize as a living thing. Which of these steps have been duplicated experimentally?

A   Life did not appear on earth until about three billion years ago. This was some two billion years after the formation of the earth, either from a portion broken off from the sun or by the gradual condensation of interstellar dust. The primitive atmosphere before the appearance of any form of life is believed to have contained essentially no free oxygen; all the oxygen atoms present were combined as water or as oxides. Deprived of free oxygen, it was thus a strongly reducing environment composed of methane, ammonia, and water which originated from the earth's interior. At that time there were obviously no organic compounds on earth.

Reactions by which organic substances can be synthesized from inorganic ones are now well known. Originally, the carbon atoms were present mainly as metallic carbides. These could have reacted with water to form acetylene, which could subsequently have polymerized to form larger organic compounds. That such reactions occurred was suggested by Melvin Calvin's experiment in which solutions of carbon dioxide and water were energetically irradiated, and formic, oxalic, and succinic acids were produced. These organic acids are important because they are intermediates in certain metabolic pathways of living organisms.

After the appearance of organic compounds, it is believed, simple amino acids evolved. How this came about was demonstrated by Urey and Miller, who in 1953 exposed a mixture of water vapor, methane, ammonia, and hydrogen gases to electric charges for a week. Amino acids such as glycine and alanine resulted. The earth's crust in prebiotic times probably contained carbides, water vapor, methane, ammonia, and hydrogen gases. Ultraviolet radiation or lightning discharges could have provided energies analogous to the Urey-Miller apparatus, and in this manner, simple organic compounds could have been produced.

Most, if not all, of the reactions by which the more complex organic substances were formed probably occurred in the sea, in which the inorganic precursors and organic products of the reaction were dissolved and mixed. These molecules collided, reacted, and aggregated in the sea to form new molecules of increasing size and complexity. Intermolecular attraction provided the means by which large, complex, specific molecules could have formed spontaneously. Once protein molecules had been formed, they acted as enzymes to catalyze other organic reactions, speeding up the rate of formation of additional molecules. As evolution progressed, proteins catalyzed the polymerization of nucleic acids, giving rise to complex DNA molecules, the hereditary materials and regulators of important functions in living organisms. Enzymes also probably catalyzed the structural combination of proteins and lipids to form membranes, permitting the accumulation of some molecules and the exclusion of others. With DNA and a membrane structure, the stage was set for life to begin some three billion years ago.

13.2 Reproduction of Organisms

A) **Asexual Reproduction** – This type of reproduction involves only a single parent and no specialized sex cell.

1) **Binary Fission** – In this type of asexual reproduction, a parent cell divides to form two new daughter cells the same size as the parent and with the same kind and quantity of DNA. This is common in protozoans, bacteria, and some species of algae, as well as in animals such as the planaria.

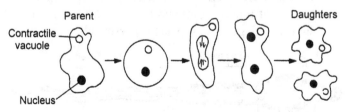

Figure 13.1 Binary fission in the amoeba

2) **Budding** – Budding is a type of asexual reproduction in which an outgrowth on a yeast cell grows and separates from the parent cell and functions as an individual. It also occurs in the hydra and its relatives.

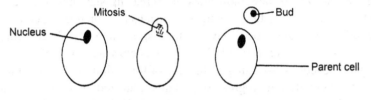

Figure 13.2 Budding in a yeast cell

3) **Spore Formation** – In this process, a cell undergoes a series of cell divisions inside its own cell wall. Each new cell produces a thick wall of its own and the old cell wall becomes a spore case (sporangium) holding small, thick walled cells, the spores. These are liberated and they grow into new individuals.

4) **Fragmentation** – This type of asexual reproduction is characterized by regeneration of any fragmented structure of a particular plant or animal.

B) **Sexual Reproduction** – In sexual reproduction, a new individual is produced by the union of two different gametes.

1) **Gonads** – The gonads are the glands which produce gametes. A testis is a male gonad; an ovary is a female gonad.

2) **Sperm Cells** – The sperm cells are the male gametes. They consist of a head, a middle piece, and a tail piece or flagellum. The flagellum propels the sperm cell.

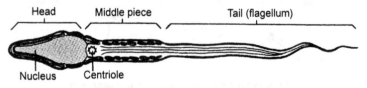

Figure 13.3 A male sperm cell

3) **Egg Cells** – The egg is the female gamete. It is larger than a sperm cell and its cytoplasm contains yolk which provides energy for the zygote divisions and the embryo's growth.

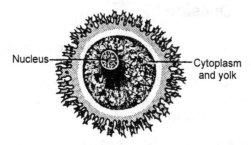

Figure 13.4 An egg cell

4) **Zygote** – A zygote is a fertilized egg formed by the union of two gametes.

Problem Solving Example:

 Differentiate between fission, budding, and fragmentation as means of asexual reproduction.

 Most protozoans reproduce asexually by fission, which is the simplest form of asexual reproduction. Fission involves the splitting of the body of the parent into two approximately equal parts, each of which becomes an entire, new, independent organism. In this case, the cell division involved is mitotic.

Hydras and yeasts reproduce by budding, a process in which a small part of the parent's body separates from the rest and develops into a new individual. It may split away from the parent and take up an independent existence or it may remain attached and maintain an independent yet colonial existence.

Lizards, starfish, and crabs can grow a new tail, leg, or arm if the original one is lost. In some cases, this ability to regenerate a missing part occurs to such an extent that it becomes a method of reproduction. The body of the parent may break into several pieces, after which each piece regenerates its respective missing parts and develops into a whole animal. Such reproduction by fragmentation is common among the flatworms, such as the planaria.

13.3 Cell Division

A) **Mitosis** – Mitosis is a form of cell division whereby each of two daughter nuclei receives the same chromosome complement as the parent nucleus. All kinds of asexual reproduction are carried out by mitosis; it is also responsible for growth, regeneration, and cell replacement in multicellular organisms.

1) **Interphase** – Interphase is no longer called the resting phase because a great deal of activity occurs during this phase. In the cytoplasm, oxidation and synthesis reactions take place. In the nucleus, DNA replicates itself and forms messenger RNA, transfer RNA, and ribosomal RNA.

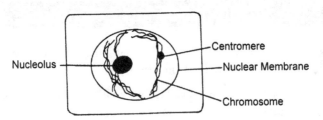

Figure 13.5 Interphase

2) **Prophase** – Chromatids shorten and thicken during this stage of mitosis. The nucleoli disappear and the nuclear membrane breaks down and disappears as well. Spindle fibers begin to form. In an animal cell, there is also division of the centrosome and centrioles.

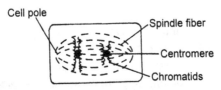

Figure 13.6 Late prophase in plant cell mitosis

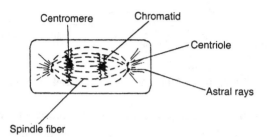

Figure 13.7 Prophase in animal cell mitosis

3) **Metaphase** – During this phase, each chromosome moves to the equator, or middle of the spindle. The paired chromosomes attach to the spindle at the centromere.

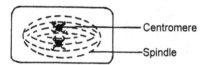

Figure 13.8 Metaphase in plant cell mitosis

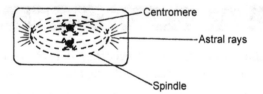

Figure 13.9 Metaphase in animal cell mitosis

4) **Anaphase** – Anaphase is characterized by the separation of sister chromatids into a single-stranded chromosome. The chromosomes migrate to opposite poles of the cell.

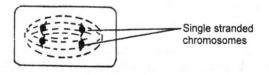

Figure 13.10 Anaphase in plant cell mitosis

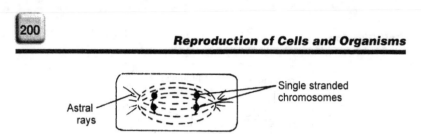

Figure 13.11 Anaphase in animal cell mitosis

5) **Telophase** – During telophase, the chromosomes begin to uncoil, and the nucleoli as well as the nuclear membrane reappear. In plant cells, a cell plate appears at the equator which divides the parent cell into two daughter cells. In animal cells, an invagination of the plasma membrane divides the parent cell.

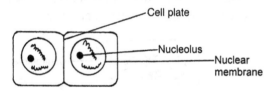

Figure 13.12 Late telophase in animal cell

Problem Solving Example:

 Outline briefly the events occurring in each stage of mitosis. Illustrate your discussion with diagrams if necessary.

Mitosis refers to the process by which a cell divides to form two daughter cells, each with exactly the same number and kind of chromosomes as the parent cell. In a strict sense, mitosis refers to the division of nuclear material (karyokinesis). Cytokinesis is the term used to refer to the division of the cytoplasm. Although each cell division is a continuous process, in order for it to be studied, it can be artificially divided into a number of stages. We will describe each stage separately, beginning with interphase.

1) Interphase: This phase is called the resting stage. However, the cell is "resting" only with respect to the visible events of division in later phases. During this phase, the nucleus is metabolically very active and chromosomal duplication is occur-

ring. During interphase, the chromosomes appear as vague, dispersed, thread-like structures and are referred to as chromatin material.

2) Prophase: Prophase begins when the chromatin threads begin to condense and appear as a tangled mass of threads within the nucleus. Each prophase chromosome is composed of two identical members resulting from duplication in interphase. Each member of the pair is called a chromatid. The two chromatids are held together at a dark, constricted area called the centromere. At this point the centromere is a single structure.

The above events occur in the nucleus of the cell. In the cytoplasm, the centriole (a cytoplasmic structure involved in division) divides and the two daughter centrioles migrate to opposite sides of the cell. From each centriole then extends a cluster of raylike filaments called an aster. Between the separating centrioles, a mitotic spindle forms, composed of protein fibrils with contractile properties. In late prophase the chromosomes are fully contracted and appear as short, rod-like bodies. At this point, individual chromosomes can be distinguished by their characteristic shapes and sizes. They migrate and line up along the equatorial plane of the spindle. Each doubled chromosome appears to be attached to the spindle at its centromere. The nucleolus (spherical body within the nucleus where RNA synthesis is believed to occur) has been undergoing dissolution during prophase. In addition, the nuclear envelope breaks down, and its disintegration marks the end of prophase.

3) Metaphase: When the chromosomes have all lined up along the equatorial plane, the dividing cell is in metaphase. At this time, the centromere divides and the chromatids become completely separate daughter chromosomes. The division of the centromeres occurs simultaneously in all the chromosomes.

4) Anaphase: The beginning of anaphase is marked by the movement of the separated chromatids (or daughter chromosomes) to opposite poles of the cell. It is thought that the chromosomes are pulled as a result of contraction of the spindle fibers in the presence of ATP. The chromosomes moving toward

the poles usually assume a V shape, with the centromere at the apex pointing toward the pole.

5) Telophase: When the chromosomes reach the poles, telophase begins. The chromosomes relax, elongate, and return to the resting condition in which only chromatin threads are visible. A nuclear membrane forms around each new daughter nucleus. This completes karyokinesis, and cytokinesis follows.

The cytoplasmic division of animal cells is accomplished by the formation of a furrow in the equatorial plane. The furrow gradually deepens and separates the cytoplasm into daughter cells, each with a nucleus. In plants, this division occurs by the formation of a cell plate, a partition which forms in the center of the spindle and grows laterally outwards to the cell wall. After the cell plate is completed, a cellulose cell wall is laid down on either side of the plate, and two complete plant cells form.

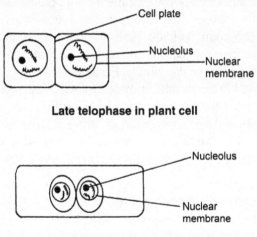

Late telophase in plant cell

Late telophase in animal cell

B) **Meiosis** – Meiosis is a second type of cell division that consists of two successive cell divisions with only one duplication of chromosomes. This results in daughter cells with a haploid number of chromosomes or one-half of the chromosome number in the original cell. This process occurs during the formation of gametes and in spore formation in plants.

1) **Spermatogenesis** – This process in sperm cell formation results in four immature sperm cells with a haploid number of chromosomes.

2) **Oogenesis** – This process results in egg cell formation with only one immature egg cell with a haploid number of chromosomes and a polar body which contains only cytoplasm.

C) First Meiotic Division

1) **Interphase I** – Chromosome duplication begins.

2) **Prophase I** – During this phase, the chromosomes shorten and thicken and synapsis occurs with pairing of homologous chromosomes. Crossing over between non-sister chromatids will also occur. The centrioles will migrate to opposite poles, and the nucleolus and nuclear membrane begin to dissolve.

3) **Metaphase I** – The tetrads, composed of two doubled homologous chromosomes, migrate to the equatorial plane.

4) **Anaphase I** – During this stage, the paired homologous chromosomes separate and move to opposite poles of the cell. Thus, the number of chromosome types in each resultant cell is reduced to the haploid number.

5) **Telophase I** – Cytoplasmic division occurs during telophase I. The formation of two new nuclei with half the chromosomes of the original cell occurs.

6) **Prophase II** – The centrioles divide, and a new spindle forms in each cell. The chromosomes move to the equator.

7) **Metaphase II** – The chromosomes, which are at right angles to the old spindle, are lined up at the equator of the new spindle.

8) **Anaphase II** – The centromeres divide and the daughter chromatids, now chromosomes, separate and move to opposite poles.

9) **Telophase II** – Cytoplasmic division occurs. The chromosomes gradually return to the dispersed form and a nuclear membrane forms.

Problem Solving Examples:

 In an animal with a haploid number of 10, how many chromosomes are present in

a)   a spermatogonium?
b)   in the first polar body?
c)   in the second polar body?
d)   in the secondary oocyte?

Assume that the animal is diploid.

 In solving this problem, one must keep in mind how meiosis is coordinated with spermatogenesis and oogenesis.

a)   Spermatogonia are the male primordial germ cells. These are the cells that may undergo spermatogenesis to produce haploid gametes. But until spermatogenesis occurs, a spermatogonium is diploid just like any other body cell. Since the haploid number is 10, the number of chromosomes in the diploid spermatogonium is 2 x 10 or 20 chromosomes.

b)   It is essential to remember that while the polar body is formed as a result of unequal distribution of cytoplasm in meiosis, the chromosomes are still distributed equally between the polar body and the oocyte. Since the first polar body is a product of the first meiotic division, it contains only one of the chromosomes of each homologous pair, since separation of homologous chromosomes has occurred. But daughter chromatids of each chromosome have not separated, so there are two identical members in each chromosome. Therefore, there are ten doubled chromosomes in the first polar body, or 20 chromatids.

c)   The second polar body results from the second meiotic division. In this division, the duplicate copies of the haploid number of chromosomes separate, forming true haploid cells. Therefore, the chromosome number is 10.

d)   The secondary oocyte results from the first meiotic division, along with the first polar body. Since, as we have said, the chromosomes have segregated equally, the secondary oocyte has the same number of chromosomes as the first polar body, and for the same reasons. Therefore, it contains 10 doubled chromosomes or 20 chromatids.

 Compare the events of mitosis with the events of meiosis. Consider chromosome duplication, centromere duplication, cytoplasmic division, and homologous chromosomes in making the comparisons.

In mitosis, the chromosomes are duplicated once, and the cytoplasm divides once. In this way, two identical daughter cells are formed, each with the same chromosome number as the mother cell. In meiosis, however, the chromosomes are duplicated once, but the cytoplasm divides two times, resulting in four daughter cells having only half the diploid chromosomal complement. This difference arises from the fact that there is no real interphase, and thus no duplication of chromosomal material between the two meiotic divisions.

In mitosis, there is no pairing of homologous chromosomes in prophase as there is in meiosis. Identical chromatids joined by their centromere are separated when the centromere divides. In meiosis, duplicated homologous chromosomes pair, forming tetrads. The daughter chromatids of each homolog are joined by a centromere as in mitosis, but it does not split in the first meiotic division. The centromeres of each duplicated member of the homologous pair are joined in the tetrad, and it is these centromeres which separate from one another in anaphase of meiosis I. Thus the first meiotic division results in two haploid daughter cells, each having chromosomes composed of two identical chromatids. Only in meiosis II, after the reduction division has already occurred, does the centromere joining daughter chromatids split as in mitosis, thus separating identical chromosomes.

Quiz: Reproduction of Cells and Organisms

1. Which of the following is not true of a human spermatid?

    (A) It has mitochondria.

    (B) It has the same number of chromosomes as a human egg.

    (C) It has half the number of chromosomes as a human egg.

    (D) It has cytoplasm.

2. Which is not a result of fertilization?

    (A) Stimulation of zygotic cleavage

    (B) Genetic determination of sex

    (C) Increase in species variability

    (D) None of the above

3. If fraternal twins arise from two separate eggs that are fertilized by two separate sperm, which of the following is not true?

    (A) They result from the simultaneous release of two eggs.

    (B) They are usually siblings of the same sex.

    (C) They have approximately 25% of their genetic information in common.

    (D) They have separate fetal membranes during their development.

4. In mitosis, gene duplication takes place during

    (A) telophase.

    (B) anaphase.

(C) interphase.

(D) metaphase.

5. If the cells of a diploid organism have 38 chromosomes, then

    (A) during mitosis, they will temporarily have 76 chromosomes.

    (B) during mitosis, they will temporarily have 19 chromosomes.

    (C) during meiosis, they will end up with 76 chromosomes.

    (D) during meiosis, they will end up with 38 chromosomes.

6. Which of the following is not a part of a prophase chromosome?

    (A) centromere

    (B) centrosome

    (C) chromatid

    (D) DNA

7. Human body cells usually have 46 chromosomes. During the anaphase stage of mitosis, a cell will have

    (A) 92 chromosomes.

    (B) 46 chromosomes.

    (C) 23 chromosomes.

    (D) 44 chromosomes.

8. Gametes are produced by

    (A) crossing over.

    (B) mitosis.

    (C) meiosis.

    (D) conjugation.

9. The primary oöcyte, which is created from oögenesis, remains in what phase of meiosis from birth to puberty?

    (A) Prophase I

    (B) Prophase II

    (C) Metaphase II

    (D) Metaphase I

10. How many gametes are generated from a single cell in spermatogenesis?

    (A) 1

    (B) 2

    (C) 3

    (D) 4

ANSWER KEY

1.	(C)	6.	(B)
2.	(D)	7.	(A)
3.	(B)	8.	(C)
4.	(C)	9.	(A)
5.	(A)	10.	(D)

CHAPTER 14

Genetic Inheritance

14.1 Mendelian Genetics

By studying one single trait at a time in garden peas, Gregor Mendel in 1857 was able to discover the basic laws of genetics.

A) Definitions:

1) **Gene** – the part of a chromosome that codes for a certain hereditary trait.

2) **Chromosome** – a filamentous or rod-shaped body in the cell nucleus that contains the genes.

3) **Genotype** – is the genetic makeup of an organism, or the set of genes that it possesses.

4) **Phenotype** – the outward, visible expression of the hereditary makeup of an organism.

5) **Homologous Chromosomes** – chromosomes bearing genes for the same characteristics.

6) **Homozygote** – an organism possessing an identical pair of alleles on homologous chromosomes for a given character or for all given characters.

7) **Heterozygote** – an organism possessing different alleles on homologous chromosomes for a given characteristic or for all given characteristics.

8) **Crossing Over** – means that paired chromosomes may break and their fragments reunite in new combinations.

9) **Translocations** – the shifting of gene positions in chromosomes that may result in a change in the serial arrangement of genes. In general, it is the transfer of a chromosome fragment to a non-homologous chromosome.

10) **Linkage** – the tendency of two or more genes on the same chromosome to be inherited together.

Table 14.1 An abstract of the data obtained by Mendel from his breeding experiments with garden peas

Parental Characters	First Generation	Second Generation	Ratios
Yellow seeds × green seeds	all yellow	6022 yellow:2001 green	3.01:1
Round seeds × wrinkled seeds	all round	5474 round:1850 wrinkled	2.96:1
Green pods × yellow pods	all green	428 green:152 yellow	2.82:1
Long stems × short stems	all long	787 long:277 short	2.84:1
Axial flowers × terminal flowers	all axial	651 axial:207 terminal	3.14:1
Inflated pods × constricted pods	all inflated	882 inflated:299 constricted	2.95:1
Red flowers × white flowers	all red	705 red:224 white	3.15:1

The 3:1 ratio resulted from this data enabled Mendel to recognize that the offspring of each plant had two factors for any given characteristic instead of a single factor.

11) **Alleles** – a type of alternative genes that occupy a given locus on a chromosome, matching genes that can control contrasting characters.

12) **Genetic Mutation** – a change in an allele or segment of a chromosome that may give rise to an altered genotype, which often leads to the expression of an altered phenotype.

13) **Codon** – a sequence of three adjacent nucleotides that codes for a single amino acid.

B) Laws of Genetics:

1) **Law of Dominance** – Of two contrasting characteristics, the dominant one may completely mask the appearance of the recessive one.

2) **Law of Segregation and Recombination** – Each trait is transmitted as an unchanging unit, independent of other traits, thereby giving the recessive traits a chance to recombine and show their presence in some of the offspring.

3) **Law of Independent Assortment** – Each character for a trait operates as a unit and the distribution of one pair of factors is independent of another pair of factors linked on different chromosomes.

C) In 1900, Walter Sutton compared the behavior of chromosomes with the behavior of the hereditary characters that Mendel had proposed and formulated the chromosome principle of inheritance.

The Chromosome Principle of Inheritance:

1) Chromosomes and Mendelian factors exist in pairs.

2) The segregation of Mendelian factors corresponds to the separation of homologous chromosomes during the reduction division stage of meiosis.

3) The recombination of Mendelian factors corresponds to the restoration of the diploid number of chromosomes at fertilization.

4) The factors that Mendel described as passing from parent to offspring correspond to the passing of chromosomes into gametes which then unite and develop into offspring.

5) The Mendelian idea that two sets of characters present in a parent assort independently corresponds to the random separation of the two sets of chromosomes as they enter a different gamete during meiosis.

6) Sutton's chromosome principle of inheritance states that the hereditary characters, or factors, that control heredity are located in the chromosomes. By 1910, the factors of hereditary were called genes.

D) The Hardy-Weinberg Law states that in a population at equilibrium both gene and genotype frequencies remain constant from generation to generation.

Problem Solving Example:

 What are the implications of the Hardy-Weinberg Law?

 The Hardy-Weinberg Law states that in a population at equilibrium both gene and genotype frequencies remain constant from generation to generation. An equilibrium population refers to a large interbreeding population in which mating is random and no selection or other factor which tends to change gene frequencies occurs.

The Hardy-Weinberg Law is a mathematical formulation which resolves the puzzle of why recessive genes do not disappear in a population over time. To illustrate the principle, let us look at the distribution in a population of a single gene pair, A and a, where A is dominant and a is recessive. Any member of the population will have the genotype AA, Aa, or aa. If these genotypes are present in the population in the ratio of 1/4 AA : 1/2 Aa : 1/4 aa, we can show that, given random mating and comparable viability of progeny in each cross, the genotypes and gene frequencies should remain the same in the next generation. Table 1 (on the following page) shows how the genotypic frequencies of AA, Aa, and aa compare in the population and among the offspring. Since the genotype frequencies are identical, it follows that the gene frequencies are also the same. It is very important to realize that the Hardy-Weinberg law is theoretical in nature and holds true only when factors which tend to change gene frequencies are absent. Examples of such factors are natural selection, mutation, migration, and genetic drift.

The Offspring of the Random Mating of a Population Composed of 1/4 AA, 1/2 Aa, and 1/4 aa Individuals

Mating Male	Female	Frequency	Offspring		
AA ×	AA	1/4 × 1/4	1/16 AA		
AA ×	Aa	1/4 × 1/2	1/16 AA +	1/16 Aa	
AA ×	aa	1/4 × 1/4		1/16 Aa	
Aa ×	AA	1/2 × 1/4	1/16 AA +	1/16 Aa	
Aa ×	Aa	1/2 × 1/2	1/16 AA +	1/8 Aa +	1/16 aa
Aa ×	aa	1/2 × 1/4		1/16 Aa +	1/16 aa
aa ×	AA	1/4 × 1/4		1/16 Aa	
aa ×	Aa	1/4 × 1/2		1/16 Aa +	1/16 aa
aa ×	aa	1/4 × 1/4			1/16 aa

Sum: 4/16 AA + 8/16 Aa + 4/16 aa

14.2 Sex Determination and X-Linkage

A) Female cells contain two identical sex chromosomes called X chromosomes. Males have only one X chromosome and a smaller Y chromosome. Both men and women have 22 pairs of autosomes, or non-sex chromosomes.

Female = XX   Male = XY

By using the Punnett square, we can understand how sex is determined by tracing the sex chromosomes to the next generation.

Possible Fertilizations

egg	sperm	
	X	Y
X	XX	XY

F₁ Offspring

Sex chromosomes	Phenotypes
1/2 XX	50% female
1/2 XY	50% male

Figure 14.1 Sex Determination – Punnett Square

B) The presence of the Y chromosome determines the maleness of an individual in humans. Its absence determines that the individual will be female.

C) **Sex-linked Traits** – Sex-linked traits are traits controlled by genes located on the X chromosome. Males tend to have sex-linked traits because they are usually recessive and males have only one X chromosome. Examples of sex-linked traits in males are hemophilia and color blindness.

D) **Chromosome Disorders** – When individuals have a chromosome number greater or fewer than the normal 46 chromosomes, certain defects may develop.

1) **Down's Syndrome** – The presence of 47 chromosomes instead of 46 makes the individual mentally retarded. The extra chromosome results from non-disjunction during the formation of the egg cell.

2) **Turner's Syndrome** – The absence of one chromosome makes this individual a short, sterile female having underdeveloped ovaries and breasts. Non-disjunction in meiosis results in an offspring lacking a sex chromosome (45 chromosomes – 44 + X).

3) **Klinefelter's Syndrome** – A male is born possessing 47 chromosomes (44 + XXY) making him tall and sterile with underdeveloped testes. This is also the result of non-disjunction of the sex chromosomes.

Problem Solving Examples:

 Explain the mechanism of the genetic determination of sex in humans.

 The sex chromosomes are an exception to the general rule that the members of a pair of chromosomes are identical in size and shape and carry allelic pairs. The sex chromosomes are not homologous chromosomes. In humans, the cells of females contain two similar sex chromosomes or X chromosomes. In males there is only one X chromosome and a smaller Y chromosome with which the X pairs during meiotic synapsis. Men have 22 pairs of ordinary chromosomes (autosomes) plus one X and one Y chromosome, and women have 22 pairs of autosomes plus two X chromosomes.

Thus, it is the presence of the Y chromosome which determines that an individual will be male. Although the mechanism is quite complex, we know that the presence of the Y chromosome stimulates the sex-organ forming portion of the egg to develop into male gonads. In the absence of the Y chromosome, and in the presence of two X chromosomes, the medulla develops into female gametes. [Note, that a full complement of two X chromosomes are needed for normal female development.]

In humans, since the male has one X and one Y chromosome, two types of sperm, or male gametes, are produced during spermatogenesis (the process of sperm formation, which includes meiosis). One half of the sperm population contains an X chromosome and the other half contains a Y chromosome. Each egg, or female gamete, contains a single X chromosome. This is because a female has only X chromosomes, and meiosis produces only gametes with X chromosomes. Fertilization of the X-bearing egg by an X-bearing sperm results in an XX, or female offspring. The fertilization of an X-bearing egg by a Y-bearing sperm results in an XY, or male offspring. Since there are approximately equal numbers of X- and Y- bearing sperm, the numbers of boys and girls born in a population are nearly equal.

 Explain how a change of one base on the DNA can result in sickle-cell hemoglobin rather than normal hemoglobin.

 If one base within a portion of DNA comprising a gene is altered, and that gene is transcribed, the codon for one amino acid on the mRNA will be different. The polypeptide chain which is translated from that mRNA may then have a different amino acid at this position. Although there are hundreds of amino acids in a protein, a change of only one can have far-reaching effects.

The human hemoglobin molecule is composed of two halves of protein, each half formed by two kinds of polypeptide chains, namely an alpha chain and a beta chain. In persons suffering from sickle-cell anemia, a mutation has occurred in the gene which forms the beta chain. As a result, a single amino acid substitution occurs, where glutamic acid is replaced by valine. One codon for glutamic acid is AUG, while one for valine is UUG. (A = adenine, U = uracil, G = guanine). If U is substituted for A in the codon for glutamic acid, the codon will specify valine. Thus, it is easy to see how a change in a single base has resulted in this amino acid substitution. This change is so important, however, that the entire hemoglobin molecule behaves differently. When the oxygen level in the blood drops, the altered molecules tend to form end-to-end associations, and the entire red blood cell is forced out of shape, forming a sickle-shaped body. These may aggregate to block the capillaries. More importantly, these sickled cells cannot carry oxygen properly and a person having them suffers from severe hemolytic anemia, which usually leads to death early in life. The condition is known as sickle-cell anemia.

In a normal hemoglobin molecule, the glutamic acid residue in question, being charged, is located on the outside of the hemoglobin molecule where it is in contact with the aqueous medium. When this residue is replaced by valine, a nonpolar residue, the solubility of the hemoglobin molecule decreases. In fact, the hydrophobic side chain of valine tends to form weak van der Waals bonds with other hydrophobic side chains leading to the aggregation of hemoglobin molecules.

14.3 DNA: The Basic Substance of Genes

A) Deoxyribonucleic acid, or DNA, is the genetic material of living organisms.

A series of experiments proved that DNA was the hereditary material. The first such evidence resulted from the transformation experiments of Fred Griffith in 1928, involving strains of pneumococcus.

DNA is the transforming principle; therefore, it is the hereditary material. Strong evidence that DNA is the genetic material came

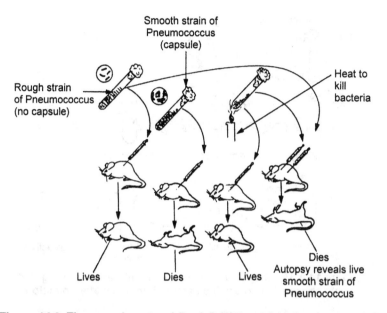

Figure 14.2 The experiments of Fred Griffith, which demonstrated the transfer of genetic information from dead, heat-killed bacteria to living bacteria of a different strain. Although neither the rough strain of Pneumococcus nor heat-killed smooth strain pneumococci would kill a mouse, a combination of the two did. Autopsy of the dead mouse showed the presence of living, smooth strain pneumococci.

from the Hershey and Chase experiments with *E. coli* and the virus which attacks *E. coli*.

B) **Transduction** – A virus transfers DNA from one bacterium to another. The work of Lederberg and Zinder (1952) with two strains of bacteria, one resistant, the other susceptible to a particular virus, provided evidence for this process.

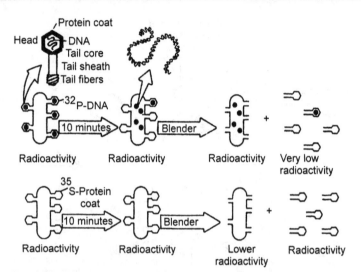

Figure 14.3 Hershey and Chase experiment demonstrating that only phage DNA enters the bacterial host cell after infection

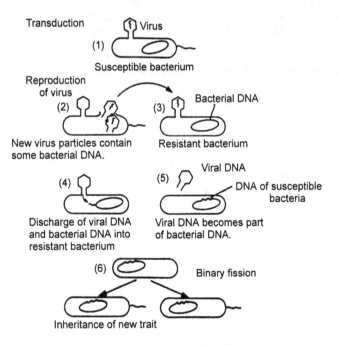

Figure 14.4 Transduction

C) The Chemical Composition of DNA:

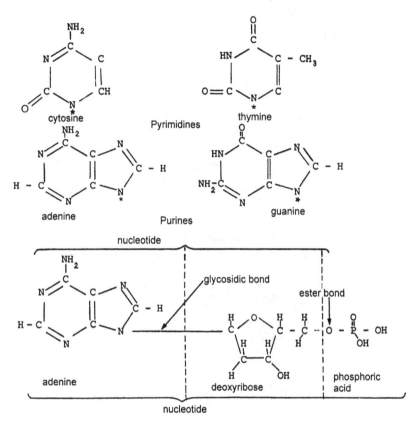

Figure 14.5 Structural formulas of purines (adenine and guanine),
pyrimidines (thymine and cytosine), and a nucleotide

Deoxyribonucleic acid is made up of a nitrogenous base, a five
carbon sugar, deoxyribose, and phosphate groups. DNA may contain one of four nitrogenous bases which are the purines, adenine
and guanine, and the pyrimidines, cytosine and thymine. Each nitrogenous base is attached to deoxyribose via a glycosidic linkage,
and deoxyribose is attached to the phosphate group by an ester
bond. This base sugar-phosphate combination is a nucleotide. There

are four kinds of nucleotides in DNA, each containing one of the four nitrogenous bases. The nucleotides are joined by phosphate ester bonds into a chain. DNA is made up of two complementary chains of nucleotides.

D) The Structure of DNA:

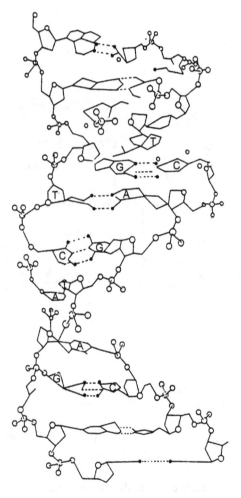

Figure 14.6 The DNA double helix

The purine (G = Guanine, A = Adenine) and pyrimidine (T = Thymine, C = Cytosine) base pairs are connected by hydrogen bonds. When these bonds are broken, the DNA can unwind and replicate (as in mitosis) or act as a template for mRNA synthesis. The ribose sugar and phosphate groups act as a backbone, helping to maintain the sequential order of the nitrogenous bases. There are ten base pairs in one turn.

Problem Solving Examples:

How does the Watson-Crick model of DNA account for its observed properties?

By 1950, several properties of DNA were well established. Chargaff showed that the four nitrogenous bases of DNA did not occur in equal proportions; however, the total amount of purines equaled the total amount of pyrimidines (A + G = T + C). In addition, the amount of adenine equaled the amount of thymine (A = T), and likewise for guanine and cytosine (G = C). Pauling had suggested that the structure of DNA might be some sort of an α-helix held together by hydrogen bonds. The final observation was made by Franklin and Wilkins. They inferred from x-ray diffraction studies that the nucleotide bases (which are planar molecules) were stacked one on top of the other like a pile of saucers.

On the basis of these observed properties of DNA, Watson and Crick in 1953 proposed a model of the DNA molecule. The Watson and Crick model consisted of a double helix of nucleotides in which the two nucleotide helices were wound around each other. Each full turn of the helix measured 34 Å (Angstroms) and contained 10 nucleotides equally spaced from each other. The radius of the helix was 10 Å. These measurements were in agreement with those obtained from X-ray diffraction patterns of DNA.

To account for Pauling's observation, Watson and Crick proposed that the sugar-phosphate chains of DNA should be on the outside, and the purines and pyrimidines on the inside, held together by hydrogen bonds between bases on opposite chains. When they tried to put the two chains of the helix together, they found the chains fitted best when they ran in opposite directions. Moreover, because X-ray diffraction studies specified the diameter of the helix to be 20 Å, the space could only accommodate one purine and one pyrimidine. If two purine bases paired, they would be too large to fit into the helix without destroying the regular shape of the double stranded structure. Similarly, two pyrimidines would be too small. Moreover, for one purine to be hydrogen-bonded with one pyrimidine properly, adenine must pair with thymine and guanine with cytosine. An A-T base pair is almost exactly the same in width as a C-G base pair, accounting for the regularity of the helix as seen in X-ray diffraction pictures. This concept of specific base pairing explained Chargaff's observation that the amounts of adenine and thymine in any DNA molecule are always equal and the amounts of guanine and cytosine are always equal. Two hydrogen bonds can form between adenine and thymine and three hydrogen bonds between guanine and cytosine. The specificity of the kind of hydrogen bond that can be formed provides for correct base pairing during replication and transcription.

The two chains are thus complementary to each other; that is, the sequence of nucleotides in one chain dictates the sequence of nucleotides in the other. The strands are also antiparallel, i.e., they extend in opposite directions and have their terminal phosphate groups at opposite ends of the double helix.

 Describe the three major structural distinctions between DNA and RNA.

 DNA or deoxyribonucleic acid is in the form of a double helix having a deoxyribose sugar and phosphate backbone. The two helices are linked together by hydrogen bonds between nitrogenous bases bound to the sugar moiety of the backbone. In DNA, the bases are adenine, guanine, cytosine and thymine.

RNA, or ribonucleic acid, differs from DNA in three important respects. First, the sugar in the sugar-phosphate backbone of RNA is ribose rather than deoxyribose. Ribose has hydroxyl group on the number 2 or 3 carbons, whereas deoxyribose has a hydroxyl on the number 3 carbon only.

Secondly, RNA has only a single sugar-phosphate backbone with at-tached single bases. Although hydrogen bonding may occur between the bases of a single RNA strand causing it to fold back on itself, it is not regular like that of DNA, where each base has a complement on the other strand. Thus, RNA is similar to a single strand of DNA. Fi-nally, the pyrimidine base uracil is found in RNA instead of the pyri-midine base thymine found in DNA.

Therefore, the nitrogenous bases present in RNA are adenine, gua-nine, cytosine and uracil.

14.4 Gene Expression

A) **RNA** is made up of a nitrogenous base, a five carbon sugar (ri-bose), and a phosphate group. RNA may contain one of four ni-trogenous bases, the purines, adenine and guanine, and the pyrim-idines, uracil and cytosine.

B) **Types of RNA** – The three types of RNA are all single-stranded and are transcribed from a DNA template by RNA polymerase in the nucleus.

   1) Messenger RNA (mRNA) carries the genetic information coded for in the DNA to the ribosomes and is responsible for the translation of that information into a polypeptide chain.

   2) Ribosomal RNA (rRNA) is an integral part of the ribosome and its removal results in the destruction of the ribosome. rRNA interacts with the ribosomal protein and helps maintain the characteristic shape of the ribosome.

   3) Transfer RNA (tRNA) is the smallest type of RNA. The func-tion of tRNA is to insert the amino acid specified by the codon on mRNA into the polypeptide chain, and it is through the complementation of anticodon and codon that the appropri-ate amino acid is incorporated.

C) The Genetic Code:

Table 14.2 The genetic code

First Position (5' end)	Second Position	Third Position (3' end) U	C	A	G
U	U	Phe	Phe	Leu	Leu
	C	Ser	Ser	Ser	Ser
	A	Tyr	Tyr	Terminator	Terminator
	G	Cys	Cys	Terminator	Trp
C	U	Leu	Leu	Leu	Leu
	C	Pro	Pro	Pro	Pro
	A	His	His	Glu	Glu
	G	Arg	Arg	Arg	Arg
A	U	Ileu	Ileu	Ileu	Met
	C	Thr	Thr	Thr	Thr
	A	Asp	Asp	Lys	Lys
	G	Ser	Ser	Arg	Arg
G	U	Val	Val	Val	Val
	C	Ala	Ala	Ala	Ala
	A	Asp	Asp	Asp	Asp
	G	Gly	Gly	Gly	Gly

The fundamental characteristic of the genetic code is that it is a triplet code with three adjacent nucleotide bases, termed a codon, specifying each tRNA and each corresponding amino acid. "Terminator" in the table refers to a codon that causes the replication.

Problem Solving Example:

Q How can autoradiography be used to show that cells without nuclei do not synthesize RNA?

A A valuable technique for tracing events in cells is autoradiography. This method relies on the fact that radioactive precursors (metabolic forerunners) taken up by the cells are incorporated into macromolecules when these cells are grown in a radioactive medium. The most widely used radioactive precursor is tritiated thymine. This

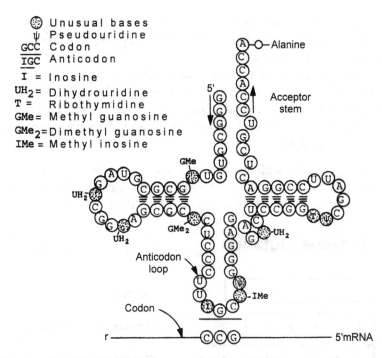

The complete nucleotide sequence of alanine tRNA showing the unusual bases and codon/anticodon position. Structure shown is two-dimensional.

is thymine (used by the cell to synthesize DNA) made radioactive by the substitution of tritium [³H], a radioactive form of hydrogen, for some of its hydrogen atoms. Typically, radioactive precursors are presented to cells. The cells are then sliced into thin strips called sections. The sections are coated with photographic emulsion similar to ordinary camera film. Exposure of camera film to light followed by photographic development leads to reduction of exposed silver salts in the film and to production of metallic silver grains. These grains form the image in the negative. Similarly, exposure to radioactivity and subsequent development produces grains in the autoradiographic emulsion. By examining the sections with an electron microscope, both the underlying structure and the small grains in the emulsion are seen.

Autoradiography can be used to show that cells without nuclei do not synthesize RNA. This can be shown by growing enucleated cells in a medium containing a radioactive precursor for RNA. One such precursor is tritiated uracil, because uracil is the pyrimidine base found only in RNA. Tritiated uracil can be made by substituting the hydrogen atoms of uracil with tritium atoms. When we make sections of enucleated cells grown in tritiated uracil and examine them under the electron microscope, we observe no grains in the sections. This is because no RNA is produced, and therefore, no radioactive precursors are incorporated into the cell.

However, cells with nuclei, when treated in a similar manner, exhibit grains in their sections. This shows that normal cells do make RNA, whereas cells without nuclei do not synthesize RNA.

14.5 Gene Mutation

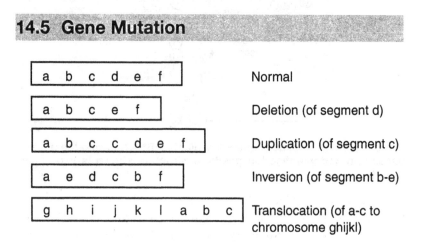

Figure 14.7 Diagram illustrating the type of mutations that involve changes in the structure of the chromosome

A) A mutation can be defined as any inheritable change in a gene not due to segregation or to the normal recombination of genetic material. There are two major types of mutation:

1) **Chromosomal Mutation** – caused by extensive chemical change in the structure of a chromosome.

2) **Point Mutation** – caused by a single change in molecular structure at a given locus.

B) Types of chromosomal mutations:

1) **Deletion** – A mutation in which a segment of the chromosome is missing.

2) **Duplication** – A mutation where a portion of a chromosome breaks off and is fused onto the homologous chromosome.

3) **Translocation** – A mutation where segments of two non-homologous chromosomes are exchanged.

4) **Inversion** – A mutation where a segment is removed and re-inserted in the same location, but in the opposite direction.

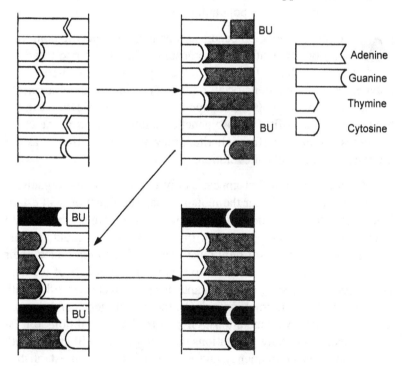

Figure 14.8 Diagrammatic scheme of how an analogue of a purine or pyrimidine might interfere with the replication process and cause a mutation, an altered sequence of nucleotides in the DNA (indicated in black). The nucleotides of the new chain at each replication are indicated by the gray blocks. In this instance, two new G-C pairs are indicated.

C) Point mutations usually involve the substitution of one nucleotide for another, and the deletion of nucleotides from the sequence and their addition to the sequence. Point mutations can result from exposure to x-rays, gamma rays, ultraviolet rays and other types of radiation, from errors in base pairing during replication, and from interaction with chemical mutagens.

Problem Solving Example:

 Define a genetic mutation. After a mutation has occurred in a population, what events must occur if the mutant trait is to become established in the population?

 A genetic mutation occurs as a change in a specific point (allele) or segment of a chromosome. This gives rise to an altered genotype, which often leads to the expression of an altered phenotype. Genetic mutations occur constantly, bringing about a variety of phenotypes in the population, upon which natural selection can act to choose the most fit. Because genetic mutations occur in the chromosomes and are therefore inherited, they are also referred to as the ultimate raw materials of evolution.

When a mutation first appears, only one or very few organisms in the population will bear the mutant gene. The mutation will establish itself in the population only if the mutants survive and breed with other members of the population. In other words, the mutants must be able to reproduce. Not only must they be able to reproduce, their zygotes must be viable and grow to become fertile adults of the next generation. By breeding of the mutants or mutation carriers either with each other or with normal individuals, the mutant gene can be transmitted to successive generations. Given the above conditions for establishment, over many generations the mutant gene will appear with greater and greater frequency and eventually become an established constituent of the population's gene pool.

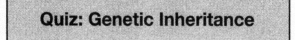

1. In an organism, pink spots is a sex-linked, recessive trait and black hair is dominant to white. If a pink spotted, black heterozygous female is mated to a white male which is not spotted, the phenotypic ratio of the male offspring would be

    (A) 1/4 pink spotted black: 1/4 pink spotted white: 1/4 unspotted black: 1/4 unspotted white.

    (B) 1/2 pink spotted black: 1/2 pink spotted white.

    (C) 3/4 pink spotted black: 1/4 pink spotted white.

    (D) 3/4 unspotted black: 1/4 pink spotted white.

2. A bb individual produces offspring with a partner of unknown genotype. Which of the following is impossible?

    (A) All of the offspring will be of phenotype B.

    (B) None of the offspring will be of phenotype B.

    (C) Some of the offspring will be heterozygous.

    (D) Some of the offspring will be homozygous dominant

3. The shape and color of radishes are controlled by two independent pairs of alleles that show no dominance. The color may be red (RR), purple (RR'), or white (R'R'), and the shape may be long (LL), oval (LL'), or round (L'L'). Red, long radishes are crossed with white, round radishes and then the F_1's are allowed to interbreed. If 1600 F_2's are obtained, then the expected ratio of white offspring would be

    (A) 400 long: 800 oval: 400 round.

    (B) 1200 long: 400 round.

(C) 50 long: 300 oval: 50 round.

(D) 100 long: 200 oval: 100 round.

4. A BbSs individual produces offspring with a bbSs individual. Which of the following would be true?

(A) About half of the offspring would show phenotype s.

(B) About 25 percent of the offspring would show phenotype b.

(C) About 25 percent of the offspring would show phenotype s.

(D) About 25 percent of the offspring would show phenotype B.

5. In peas, yellow color is dominant to green. If a homozygous yellow plant is crossed with a green plant, then the expected phenotypic ratios of the offspring would be

(A) 3/4 yellow: 1/4 green.

(B) all green.

(C) 3/4 green: 1/4 yellow.

(D) all yellow.

6. In certain types of dogs the allele for a spotted coat is recessive to that of a solid pattern. If 4% of the dogs in a given population of 10,000 are spotted, how many are expected to be heterozygous?

(A) 4,800

(B) 9,600

(C) 3,200

(D) 800

7. Each of the following is an important assumption for maintenance of a Hardy-Weinberg equilibrium in a population EXCEPT

    (A) asexual reproduction.

    (B) random mating among members.

    (C) large population size.

    (D) lack of emigration or immigration.

8. Which of the following is likely to be a mutation that has a serious consequence for the organism that experiences it?

    (A) A point mutation in which a single base A is mistakenly inserted instead of base T.

    (B) A point mutation in which a single base A is mistakenly failed to be inserted.

    (C) A point mutation in which a single base C is mistakenly inserted instead of base G.

    (D) A mutation in which two adjacent bases switch places.

9. The Hardy-Weinberg rule states that under certain conditions evolution cannot occur. Which of the following conditions is not required by the Hardy-Weinberg Law?

    (A) No mutations

    (B) No immigration or emigration

    (C) No natural selection

    (D) No isolation

10. Nondisjunction, whereby a pair of homologous chromosomes does not separate in the first meiotic anaphase, is responsible for all of the following disorders EXCEPT

    (A) Turner's Syndrome (XO).

    (B) Down's Syndrome (trisomy-21).

    (C) Klinefelter's Syndrome (XXY).

    (D) Hemophilia (X^hY or X^hX^h).

ANSWER KEY

1. (B)
2. (D)
3. (D)
4. (B)
5. (D)

6. (C)
7. (A)
8. (B)
9. (D)
10. (D)

CHAPTER 15

Embryonic Development

15.1 Stages of Embryonic Development

The earliest stage in embryonic development is the one-cell, diploid zygote which results from the fertilization of an ovum by a sperm.

A) **Cleavage** – Cleavage is a series of mitotic divisions of the zygote which result in the formation of daughter cells called blastomeres.

B) **Morula** – A morula is a solid ball of 16 blastomeres.

C) **Blastula** – A blastula is a hollow ball; a fluid-filled cavity at the center of the sphere is the blastocoel.

D) **Gastrula** – The cells of the blastula have differentiated into two, and then three embryonic germ layers, forming a gastrula. Early forms of all major structures are laid down in the gastrula period. After this period, the developing organism is called a fetus.

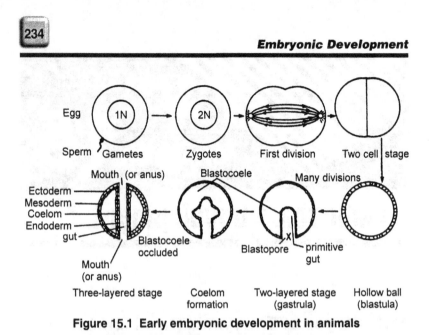

Figure 15.1 Early embryonic development in animals

Table 15.1 Derivatives of the Primary Germ Layers

Primary Germ Layer	Derivatives
Endoderm	Inner lining of alimentary canal and respiratory tract; inner lining of liver; pancreas; salivary, thyroid, parathyroid, thymus glands; urinary bladder; urethra lining
Mesoderm	Skeletal system; muscular system; reproductive system; excretory system; circulatory system; dermis of skin; connective tissue
Ectoderm	Epidermis; sweat glands; hair; nails; skin; nervous system; parts of eye, ear, and skin receptors; pituitary and adrenal glands; enamel of teeth

Problem Solving Example:

 What is meant by embryonic development? Describe the various stages of embryonic development.

 Embryonic development begins when an ovum is fertilized by a sperm and ends at parturition (birth). It is a process of change and growth which transforms a single cell zygote into a multicellular organism.

The earliest stage of embryonic development is the one-celled, diploid zygote which results from the fertilization of an ovum by a sperm. Next is a period called cleavage, in which mitotic division of the zygote results in the formation of daughter cells called blastomeres. At each succeeding division, the blastomeres become smaller and smaller. When 16 or so blastomeres have formed, the solid ball of cells is called a morula. As the morula divides further, a fluid-filled cavity is formed in the center of the sphere, converting the morula into a hollow ball of cells called a blastula. The fluid filled cavity is called the blastocoel. When cells of the blastula differentiate into two, and later three, embryonic germ layers, the blastula is called a gastrula. The gastrula period generally extends until the early forms of all major structures (for example, the heart) are laid down. After this period, the developing organism is called a fetus. During the fetal period (the duration of which varies with different species), the various systems develop further. Though developmental changes in the fetal period are not as dramatic as those occurring during the earlier embryonic periods, they are extremely important.

Congenital defects may result from abnormal development during this period.

15.2 Types of Eggs

The eggs of different animals vary greatly in the amount and distribution of the yolk they contain.

A) **Isolecithal Eggs** – The yolk is evenly distributed. Example: eggs of annelids, mollusks, and echinoderms

B) **Telolecithal Eggs** – The yolk is concentrated toward one pole of the egg, the vegetal pole. Examples: reptiles and bird eggs

C) **Centrolecithal Eggs** – The yolk is massed toward the center. Example: insect eggs

Problem Solving Example:

Q Following fertilization, the zygote begins to cleave. Describe the different cleavage patterns found in the animal kingdom.

Radial and spiral cleavage patterns. Left: Radial cleavage, characteristic of deuterostomes. The cells of the two layers are arranged directly above each other. Right: Spiral cleavage, characteristic of protostomes. The cells in the upper layer are located in the angles between the cells of the lower layer.

A Cleavage cells are known as blastomeres. They vary in size and content, principally by reason of differences in the amount and distribution of the yolk and other cytoplasmic inclusions which they contain. In isolecithal eggs, the cleavage cells are approximately the same size. In telolecithal eggs, such as those of birds, the yolk is more abundant and is concentrated toward that area of the egg known as the vegetal pole. The opposite side of the egg, an area where the nuclear material is located, is called the animal pole. In this case, the blastomeres nearer the animal pole tend to be smaller and are called micromeres. The cells near the vegetal pole are usually larger and are termed macromeres. As long as the entire egg divides into cells, the cleavage is said to be complete, or holoblastic. In the case of sharks, reptiles, and birds, however, yolk is abundant and fills the egg, except for a thin disc of cytoplasm at the animal pole and an even thinner layer of cytoplasm around the periphery of the egg. In such eggs, cleavage is incomplete and is confined to a small area which surrounds the animal pole. The rest of the egg remains uncleaved. Such incomplete cleavage is termed meroblastic. In the centrolecithal eggs of arthropods, the yolk is concentrated toward the center. After the nucleus divides several times, the offspring nuclei migrate to the periphery of the egg where meroblastic cleavage of the peripheral cytoplasm takes place.

There are three basic patterns of cleavage in the animal kingdom: radial cleavage, spiral cleavage, and superficial cleavage. In radial cleavage, the first mitotic spindle elongates in a direction at right angles to the egg axis. The second spindle also elongates transversely to the egg axis and at right angles to the first spindle. The third cleavage spindle elongates in a direction parallel to the egg axis. Consequently, a ball-like stage of eight cleavage cells is formed. This type of cleavage is found in the echinoderms and chordates. In spiral cleavage, the mitotic spindles are oblique to the polar axis of the embryo, and give rise to a spiral arrangement of newly formed cells. This type of cleavage is found in the annelids, mollusks, and in some flatworms. A third type of cleavage, associated with arthropods, is known as superficial cleavage. This involves mitotic division of the nucleus without any cleavage, resulting in an uncleaved egg containing a large number of nuclei within the center. Later, these nuclei migrate to the periphery where meroblastic cleavage of the egg takes place. The rate of cleavage in all three cases is rapid at first and then slows a bit and can be regulated by temperature changes.

15.3 Cleavage and Gastrulation

A) **Cleavage and Gastrulation of a Mammalian Egg** – Cleavage results when a solid ball of cells, the morula, becomes subdivided into an inner cell mass from which the embryo develops and an enveloping layer of cells, the trophoblast. The cells of the inner cell mass differentiate further into a thin layer of flat cells, the hypoblast. The remaining cells of the inner cell mass become the epiblast. Gastrulation begins with the formation of a primitive streak and Henson's node in which cells migrate downward, laterally, and anteriorly between the epiblast and hypoblast. A crevice appears between the cells of the inner cell mass, which then enlarges to become the amniotic cavity. The embryonic disc comes to lie as a plate between the two cavities, connected to the trophoblast. The non-functional endodermal part of the allantois develops as a tube from the yolk sac. After two weeks the human embryo is a flat, two-layered disc of cells about 250 microns across connected by a stalk to the trophoblast.

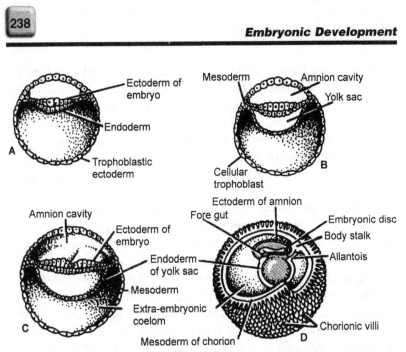

Figure 15.2 Diagrams of human embryos ten to twenty days old showing the formation of the amniotic and yolk sac cavities and the origin of the embryonic disc

B) Cleavage and Gastrulation in Other Organisms

Starfish – Cleavage is radial in the isolecithal starfish egg. The

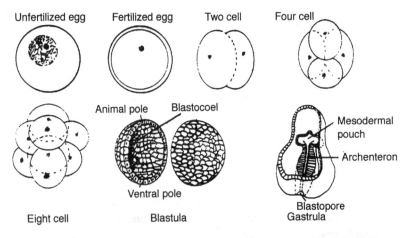

Figure 15.3 Some developmental stages in the sea star (starfish)

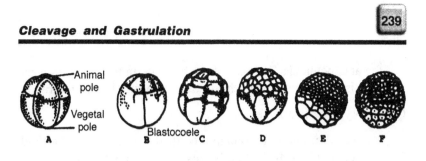

Figure 15.4 Early embryology in a frog

egg is split into two equal daughter cells. The second cleavage plane is vertical but at right angles to the first, separating the two cells into four. With further divisions, a blastula forms. It is a simple layered blastula which is converted later into a double layered sphere, the gastrula, by the invagination of a section of a wall of the blastula.

Amphibians – The cleavage pattern in amphibians is radial, but because of the unequal distribution of yolk in the telolecithal egg, the blastomeres are not of equal size. At the eight-celled stage, the egg consists of four smaller animal pole blastomeres (micromeres), and four larger, yolk-laden vegetal pole blastomeres (macromeres); the egg then becomes a gastrula. To form a gastrula, a groove appears on one side of the blastula and cells at the bottom of the groove stream into the interior of the embryo. This groove extends transversely until it is ring-shaped, at which point, the yolk-filled cells remain as a yolk plug.

Birds – Telolecithal avian eggs display meroblastic radial cleavage. Cleavage occurs in a small disc of cytoplasm in the animal pole. Horizontal cleavage separates upper blastomeres and lower blastomeres. The blastomeres at the margin of the disc, and the lower cells in contact with the yolk, lose the furrows that partially separated them and fuse into a continuous syncytium which contains many nuclei and is called the periblast. The free blastomeres become incorporated into two layers, an upper epiblast and a lower hypoblast. Between them is a blastocoele. The subgerminal space separates the hypoblast from the underlying yolk. The area of the blastoderm over the subgerminal space is transparent and is called the area pellucida. The part of the blastoderm that contains some yolk is called the area opaca. Gastrula is formed by cell migration.

C) Embryonic Membranes

1) **The Amnion** – The amnion is filled with a lymph-like fluid, the amniotic fluid, that bathes and protects the embryo in a watery environment. It surrounds the embryo.

2) **The Yolk Sac** – Many blood vessels develop in the walls of the yolk sac so that food material is transported from the yolk to the cell of the developing embryo. It grows around the yolk in the egg.

3) **The Allantois** – The allantois is a sac-like structure that covers both the embryo and the yolk sac. The capillaries of the allantois exchange carbon dioxide and oxygen with the atmosphere.

4) **The Chorion** – The chorion is the outermost embryonic membrane, forming a moist lining underneath the other membranes. The chorion helps in the exchange of respiratory gases between the shell and the capillaries of the allantois.

Problem Solving Example:

 Contrast the development of an amphibian, a bird, and a starfish egg up to the point of gastrulation.

 In the isolecithal starfish egg, cleavage is radial. The first cleavage division passes through the long axis of the egg, splitting the egg into two equal daughter cells. The second cleavage plane is also vertical but at right angles to the first, separating the two cells into four. The third division is horizontal, at right angles to the other two, and splits the four cells into eight. Further divisions result in embryos containing 16, 32, 64, 128 cells, and so on, until a hollow ball of cells, called the blastula, is formed.

The wall of the blastula is a single layer of cells, the blastoderm, surrounding the blastocoel, the cavity in the center. It is this single-layered blastula which is converted later into a double-layered sphere, the gastrula.

Amphibians also display radial cleavage, but because of the unequal distribution of yolk in the telolecithal egg, the resulting blastomeres are not of equal size. The first cleavage begins as a shallow furrow at the animal pole and progresses gradually through to the vegetal pole. This first cleavage usually results in two equal blastomeres. The second cleavage is at right angles to the first. It begins at the animal pole even before the first cleavage has been completed at the vegetal pole. The result is four blastomeres of nearly equal size. The third cleavage is typically at right angles to the first two cleavages, and hence cuts horizontally across the egg's vertical axis. It passes well above the equator, so that the eight-celled stage commonly consists of four smaller animal blastomeres (micromeres), and four larger, yolk-laden vegetal blastomeres (macromeres).

A cavity is present at the center of the group of blastomeres beginning at the eight-celled or 16-celled stage, which increases in size as cleavage progresses. The egg thus becomes a hollow ball of cells, namely, a blastula. Its blastocoel is roughly hemispherical. It has a dome of smaller animal cells and a floor of larger yolk-laden vegetal cells.

Telolecithal bird eggs also display radial cleavage, but the cleavage is meroblastic. In bird eggs and other eggs which contain a large amount of yolk, cleavage occurs only in a small disk of cytoplasm in the animal pole. At first, all cleavage planes are vertical and all the blastomeres lie in a single plane. The cleavage furrows separate the blastomeres from each other but not from the yolk. The central blastomeres are continuous with the yolk at their lower ends, and the blastomeres at the circumference of the disk are continuous both with the yolk beneath them and with the uncleaved cytoplasm at their outer edge. As cleavage continues, more cells become cut off to join the ones in the center, but the new blastomeres are also continuous with the uncleaved underlying yolk. The central blastomeres eventually become separated from the underlying yolk either by cell divisions with horizontal cleavage planes or by the appearance of slits beneath the nucleated portions of the cells. Horizontal cleavages separate upper blas-

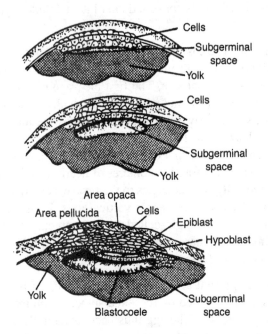

Successive stages in the cleavage of a hen's egg. Cleavage is restricted to a small disc of cytoplasm on the upper surface of the egg yolk called the blastodermic disc. A subgerminal space appears beneath the blastodermic disc, separating it from the unsegmented yolk. The blastodermic disc cleaves into an upper epiblast and a lower hypoblast separated by the blastocoel.

tomeres and lower blastomeres. The upper blastomere is a cell with a complete plasma membrane, which is separated from its neighbors and from the yolk. The lower blastomere is a cell which remains connected with the yolk. The blastomeres at the margin of the disk and the lower cells in contact with yolk eventually lose the furrows that partially separated them and fuse into a continuous syncytium. This syncytium contains many nuclei and is termed the periblast. The periblast is believed to break down the yolk, thereby making its nutrients available for the growing embryo. The free blastomeres (with complete plasma membranes) become incorporated into two layers, an upper epiblast and a lower hypoblast. Between these two layers is a cavity, the blasto-coel. Below the hypoblast and above the yolk is the shallow subgerminal cavity, which appears only under the central portion of the blastoderm. The area of the blastoderm over the subgerminal space is relatively transparent (due to lack of yolk) and is called the area pellucida, whereas the more opaque part of the blastoderm (which contains some yolk) that rests directly on the yolk is called the area opaca.

15.4 Parturition

Parturition is the separation of the fetus and its membranes from the mother's body at the time of birth. As gestation ends, the hormone relaxin appears in the bloodstream, causing the cervix of the uterus to become readily dilatable and loosening the connections between the bones of the pelvis. Oxytocin stimulates the uterine muscles to contract with greater force and frequency during labor.

Rhythmic contractions of the smooth muscles of the uterus forces the fetus, normally head first, into the cervix of the uterus. The fetal membranes then bulge through the cervix and burst. Amniotic fluid is discharged. The child then enters the vagina and is born.

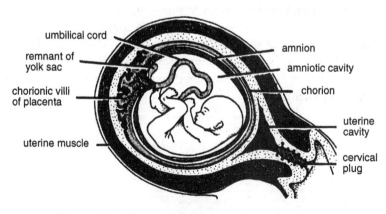

Figure 15.5 Diagram of advanced fetus showing its membranes and their relationship to the uterus

Problem Solving Example:

Q Parturition is the separation of the fetus and its membranes from the mother's body at the time of birth. Describe this process.

A As the period of gestation ends, a hormonal substance termed relaxin appears in the bloodstream. This substance seems to be a product of the corpus luteum of pregnancy and of the placenta and causes the cervix of the uterus to become readily dilatable. Oxytocin, one of the hormones released from the posterior pituitary, is an extremely potent uterine-muscle stimulant. Oxytocin is released by reflex into the bloodstream as a result of afferent input to the hypothalamus from receptors in the uterus, particularly in the cervix. Oxytocin stimulates the uterine muscles to contract with greater force and frequency during labor. Relaxin and oxytocin are two important hormones involved in parturition.

At the time of birth, rhythmic contractions of the smooth muscle of the uterus begin. Soon they are accompanied by reflex and voluntary activity of skeletal muscle. The result is that internal pressure forces the fetus, normally head first, into the cervix of the uterus. The fetal membranes then bulge through the cervix and burst. Amniotic fluid is discharged. The child then enters the vagina and is born, a process lasting from twenty minutes to one hour. Some fifteen minutes after birth, the uterus again goes into rhythmic contraction, and an "afterbirth" consisting of the fetal membranes is expelled. Surprisingly enough, there is only a minor amount of bleeding. Within the next few days, the lining of the uterus is restored and the uterus returns to nearly its original dimensions.

Quiz: Embryonic Development

1. Following fertilization, the zygote undergoes a series of rapid mitotic divisions. The stage at which a solid ball of cells is formed is called

    (A) the morula.

    (B) the blastula.

    (C) the gastrula.

    (D) the fetus.

2. The first stage of embryonic development in which three distinct germ layers are seen is

    (A) the morula.

    (B) the blastula.

    (C) the gastrula.

    (D) the fetus.

3. An egg in which the yolk is concentrated toward one pole is called

    (A) telolecithal.

    (B) centrolecithal.

    (C) isolecithal.

    (D) mesolecithal.

4. The smaller sized blastomeres are called _____, and they are found at the _____ pole.

    (A) macromeres, animal

    (B) macromeres, vegetal

    (C) micromeres, animal

    (D) micromeres, vegetal

5. Which is true concerning the process of cleavage?

    (A) Meroblastic cleavage is characteristic of human embryology.

    (B) Meroblastic cleavage results in complete cell cleavage.

    (C) Radial cleavage is characteristic of chordates.

    (D) Spiral cleavage occurs only in vertebrates.

6. Which of the following structures are not ectodermal in origin?

    (A) Kidney, ureter, gonads

    (B) Sense organs, anal canal, nasal cavity

    (C) Epidermis, skin glands, tooth enamel

    (D) Brain, spinal cord, nerves

7. Which of the following are not mesodermal in origin?

    (A) Muscle tissue, cartilage, bone

    (B) Bone marrow, blood, lymphoid tissue

    (C) Blood vessels, genital ducts, kidneys

    (D) Thymus gland, trachea, bladder

8. When a cell is in the prophase of mitosis,

   (A) the centromeres replicate and separate.

   (B) the nuclear membrane disintegrates and spindle fibers are formed.

   (C) the chromatids are pulled apart.

   (D) the separated chromatids become the chromosomes of the daughter cells that are to form.

9. A zygote will produce a 32-cell blastula after dividing mitotically by a number of divisions equalling:

   (A) 2.

   (B) 4.

   (C) 5.

   (D) 7.

10. Choose the proper sequence of development.

    (A) morula, blastula, gastrula, neurula

    (B) morula, gastrula, blastula, neurula

    (C) neurula, gastrula, blastula, morula

    (D) blastula, neurula, gastrula, morula

ANSWER KEY

1.	(A)	6.	(A)
2.	(C)	7.	(D)
3.	(A)	8.	(B)
4.	(C)	9.	(C)
5.	(C)	10.	(A)

CHAPTER 16

Evolution

16.1 Evidence for Evolution

A) Historical Development of the Theory of Evolution

1) **Jean de Lamarck** – In 1809, Lamarck proposed that, from generation to generation, acquired characteristics are inherited. Lamarck believed that new organs arise in response to demands of the environment, and that their sizes are proportional to their "use or disuse." These changes in size were believed to be inherited by succeeding generations. This theory has been rejected because of an overwhelming amount of evidence which indicates that acquired traits cannot be inherited.

2) **The Darwin-Wallace Theory of Natural Selection** – The Darwin-Wallace theory of natural selection states that a significant part of evolution is dictated by natural forces, which select for survival those organisms that can respond best to certain conditions. Since more organisms are born than can be accommodated by the environment, a limited number is chosen to live and reproduce. Variation is characteristic of all animals and plants, and it is this variety which provides the means for this choice. Those individuals who are chosen for survival will be the ones with the most and best adaptive traits. These include the ability to compete successfully for food, water, shelter, and other essential elements, also the

ability to reproduce and perpetuate the species, and the ability to resist adverse natural forces, which are the agents of selection.

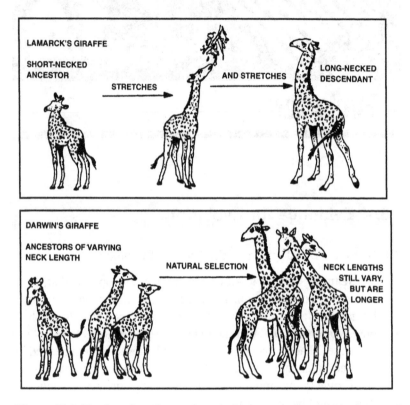

Figure 16.1 Rival explanations of evolution were advanced by Lamarck and Darwin. According to Lamarck, the long neck of the giraffe arose as a result of generations of stretching to reach food. According to Darwin, giraffes with longer necks arose as a result of natural selection.

3) **De Vries' Theory of Mutation** – De Vries proposed that variations arise as the result of spontaneous mutations that are inherited.

4) **The Modern (Neo-Darwinian) Theory of Evolution** – This theory includes overproduction, which results in a struggle for existence. The organisms that survive are those who are best adapted to the conditions of life. Only variations that result from changes

in genes, DNA, and chromosomes are inherited by the following generations.

B) Evidence for Evolution

1) **Comparative Anatomy** – Similarities of organs in related organisms show common ancestry.

2) **Vestigial Structures** – Structures of no apparent use to the organism may be explained by descent from forms that used these structures.

3) **Comparative Embryology** – The embryo goes through developmental stages in common with other types of species.

4) **Comparative Physiology** – Many different organisms have similar enzymes. Mammals have similar hormones.

5) **Taxonomy** – All organisms can be classified into kingdom, phylum, class, order, family, genus and species. This commonness in classification seems to indicate relationships between organisms.

6) **Biogeography** – Natural barriers such as oceans, deserts, and mountains restrict the spread of species to other favorable environments. Isolation frequently produces many variations of species.

7) **Genetics** – Gene mutations, chromosome segment rearrangements, and chromosome segment doubling produce variations and new species.

8) **Paleontology** – Present individual species can be traced back to origins through skeletal fossils.

Problem Solving Examples:

Describe briefly the Darwin-Wallace theory of natural selection. What is meant by "survival of the fittest"? Do you think this applies to human populations today?

The Darwin-Wallace theory of natural selection states that a significant part of evolution is dictated by natural forces which select for survival those organisms that can respond best to them. Since more organisms are born than can be accommodated by the environment, there must be selected among the large numbers born the limited number to live and reproduce. Variety provides the means for this choice. Those individuals who are selected for survival will be the ones with the most and best adaptive traits. These include the ability to compete successfully for food, water, shelter, and other essential elements. They also include the ability to reproduce and perpetuate the species and the ability to resist adverse natural forces, which are the agents of selection.

Essential to the theory of natural selection are the ideas of the "struggle for existence" and "survival of the fittest." Because the resources of the environment are naturally limited, individuals must struggle among themselves for food and space. Those individuals with the traits that are best suited for the given environment will survive and multiply, while others will slowly decline in number or disappear completely. Since the natural forces are constantly operating and changing, struggle for survival goes on forever, and the fittest of the competitors survive.

Survival of the fittest holds true for human populations, but it is less obvious among human beings than organisms. Whereas it is common to see wild dogs fighting each other for a piece of food, it is less common to see two human beings fighting for the same reason. Among human beings, the problems of food and shelter have been conquered successfully, such as by farming or by raising cattle. Actual physical struggle between human beings for food has become less important. The traits of strong muscles, great agility and alertness, and other fighting abilities that were once important to the cavemen have more or less lost their significance today.

 Describe the various types of evidence from living organisms which support the theory of evolution.

 There are five lines of evidence from living organisms that support the theory of evolution. First, there is the evidence from taxonomy. The characteristics of living things differ in so orderly a pattern that they can be fitted into a hierarchical scheme of categories. Our present, well-established classification scheme of living organisms, developed by Carolus Linnaeus in the 1750s, groups organisms into the Kingdom, Phylum (or Division), Class, Order, Family, Genus, and Species. The relationships between organisms evident in this scheme indicates evolutionary development. If the kinds of plants and animals were not related by evolutionary descent, their characteristics would most probably be distributed in a confused, random fashion, and a well-organized classification scheme would be impossible. Secondly, there is the evidence from morphology. Comparisons of the structures of groups of organisms show that their organ systems have a fundamentally similar pattern that is varied to some extent among the members of a given phylum. This is readily exemplified by the structures of the skeletal, circulatory, and excretory systems of the vertebrates. The observation of homologous organs—organs that are basically similar in their structures, site of occurrence in the body, and embryonic development, but are adapted for quite different functions—provides a strong argument for a common ancestral origin. In addition, the presence of vestigial organs, which are useless or degenerate structures found in the body, points to the existence of some ancestral forms in which these organs were once functional. Thirdly, there is the evidence from comparative biochemistry. For example, the degree of similarity between the plasma proteins of various animal groups, tested by an antigen-antibody technique, indicates an evolutionary relationship between these groups. Fourthly, embryological structures and development further support the occurrence of evolution. Different animal groups have been shown to have a similar embryological form. It is now clear that at certain stages of development, the embryos of the higher animals resemble the embryos of lower forms. The similarity in the early developmental stage of all vertebrate embryos indicates that the various vertebrate groups must have evolved from a common ancestral form. Finally, there is the evidence from genetics. Breeding experiments and results demonstrate that species are not unchangeable biologic entities which were created separately, but groups of organisms that have arisen from other species and that can give rise to still others.

16.2 Mechanisms of Evolution

Evolution is the result of the interaction of four major forces. These are mutation, genetic drift, migration, and natural selection.

1) Mutations are random events that occur at a very low rate of approximately one out of every 10^6 genes. Mutation introduces variety into a population.

2) Genetic drift acts in the evolutionary process by causing chance fluctuations in gene pool frequencies.

3) Migration occurs when individuals from one breeding population leave to join another. Migration may lead to either more variation in a population due to the introduction of new genes, or less variation due to the loss of genes through emigration.

4) Natural selection assures that the changes in allele frequency caused by migration, mutation, and genetic drift are adaptive. Natural selection operates through differential reproduction, which occurs when certain individuals are able, by surviving and/or reproducing at a higher rate, to preferentially propagate and transmit their respective genes over those of other individuals.

Problem Solving Example:

What is meant by "genetic drift"? What role may this play in evolution?

The random fluctuation of gene frequencies due to chance processes in a finite population is known as genetic drift. Genetic drift may occur by what is commonly referred to as the Founder Principle. This principle operates by chance factors alone. A group of individuals may leave their dwelling and establish a new feeding or breeding ground in an unexploited area. The individuals that find the new area may not represent the same genetic makeup as the original population, since the individuals that left were a random group. This exhibits how one population may establish itself with a new genetic composition by chance alone. Genetic drift may also affect a population without the migration of individuals. Catastrophes in one area of a certain population, such as a flood, may randomly kill individuals of a population. Those individuals which survive do so by chance alone and not by possessing any special selective advantage. Thus genetic drift may result from natural catastrophes where a random group of individuals may survive by chance alone and change the original population's allelic frequencies. The alteration of allelic frequencies by chance due to genetic drift is unlike the directional movements caused by the systematic pressures of mutation, selection, and differential migration. The situation is quite different in a large population.

In large populations, the relatively small numbers of chance variations in gene frequencies are absorbed into the population in succeeding generations and the overall effect is negligible. In small interbreeding populations, however, where there are limited numbers of progeny in a given generation, random variation has a significant effect and genetic drift can become an important factor. Random fluctuation can most often lead to fixation of a given allele in a small population.

That is, one allele is slowly replacing the other as a result of chance, and the former becomes fixed in the population as the latter is lost. This process of reduction of heterozygosity through loss and fixation at various loci is also known as the decay of variability.

The fixation of certain genes by means of genetic drift may explain the appearance of seemingly unimportant and useless structures in some organisms. These structures may be the expression of homozygous gene pairs which have accumulated in the population by chance alone. Such a loss of variability can also inactivate natural selection, which can act only when a certain degree of phenotypic variation is present. If there is a lack of variation, any adverse situation, such as a period of extreme cold or the arrival of a group of predators, could terminate an entire population. In such a homogeneous population, only genetic mutation could hopefully preserve the population. This happens as some mutants arise which are resistant to the unfavorable force. However, since the frequency of an allelic mutation is in the area of 1 in 10^6, a homogeneous population could find itself in great danger of extinction.

16.3 The Five Principles of Evolution

1) Evolution occurs more rapidly at some times than at others.

2) Evolution does not proceed at the same rate in all types of organisms.

3) Most new species do not evolve from the most advanced and specialized forms already living, but from relatively simple, unspecialized forms.

4) Evolution is not always from the simple to the complex, but may act in such a way that complex forms have given rise to simpler ones. This is termed as "regressive evolution."

5) Evolution occurs in populations (not in individuals) in a gradual process over very long periods of time.

Adaptive radiation   Convergent evolution
divergent evolution

Figure 16.2 Diagram illustrating the difference between divergent and convergent evolution. In adaptive radiation, a single stock may branch to give many diverging stocks. In convergent evolution, many stocks that are originally quite different can come to resemble each other more and more; as time passes, they converge.

Definitions:

1) **Adaptive Radiation** – A single ancestral species evolves to a variety of forms that occupy somewhat different habitats.

2) **Convergent Evolution** – Two or more unrelated groups of organisms become adapted to a similar environment and develop characteristics that are more or less similar.

Problem Solving Example:

Q From what you have learned of the processes of heredity and evolution, how do you think new species arise with respect to mutation? Which is more important, the accumulation of small mutations or a few mutations with large phenotypic effects? Give the reasons for your answer.

A Most adaptive mutations produce barely distinguishable changes in the structure and function of the organism in which they occur. On the other hand, mutations with the potential to produce large phenotypic effects necessary to elicit major changes in one jump are practically always lethal or at least detrimental to survival. Major mutations are deleterious because they constitute a great disturbance to the delicately balanced genetic systems in which they arise. Therefore, major genotypic changes through mutation are not likely to be important in evolution.

In contrast, the accumulation of small mutations over the course of many generations has significant evolutionary value. This has been supported by fossil evidence, laboratory experiments, and field studies. Evolutionary changes appear to arise as a result of many small changes that have accumulated over time, ultimately producing a change in the composition of a given population's gene pool. An example of this can be seen in the evolution of the orchid. Some species of modern orchid flowers, in both shape and color, resemble the female wasp. Male wasps, thinking it is a female, are attracted to the flower and stimulated to attempt copulation. In doing so, they become coated with the sticky pollen grains of the flower and carry them along to the next orchid. Any mutation that gave the orchid any similarity, however slight, to the female wasp would increase its chances of being pollinated. Thus, such mutations would give those orchids bearing them a selective advantage over orchids without them. The increased rate of pollination of the mutated forms would promote the propagation of the mutant gene. It is easy to see how, over time, such mutations would tend to be propagated and accumulate to the point where the orchid flower is today.

16.4 Classification

A) **Classification** – The modern day basis of the classification of living organisms was developed by Linnaeus. Linnaeus based his system of classification upon similarities of structure and function between different organisms. The bases of classifying living things are not only structure and function, but development of the young, similarity of DNA, and adaptations for survival. The classification groups are: kingdom, phylum, division, class, order, family, genus and species.

B) **Kingdom Protista** – Most protists are unicellular. Some are composed of colonies. All protists are eukaryotes. That is, they are characterized by nuclei bounded by a nuclear membrane. Examples of protists are flagellates, the protozoans and slime molds.

C) **Kingdom Eubacteria** – Bacteria and blue-green algae are included in the kingdom Eubacteria. These cells have no nuclear membrane and only a single chromosome are termed prokaryotes. Both blue-green algae and bacteria lack membrane-bounded subcellular organelles such as mitochondria and chloroplasts.

D) **Kingdom Archaebacteria** – This form of bacteria is so different from the other bacteria that it comprises a separate kingdom termed the Archaebacteria.

The other kingdoms are the Animal, Plant, and Fungi kingdoms.

Problem Solving Example:

The following are all classification groups: family, genus, kingdom, order, phylum, species, and class. Rearrange these so that they are in the proper order of sequence from the smallest grouping to the largest. Explain the scientific naming of a species.

Closely related species are grouped together into genera (singular—genus), closely related genera are grouped into families, families are grouped into orders, orders into classes, classes into phyla (singular—phylum), and phyla into kingdoms. Classes and phyla are the major divisions of the animal and plant kingdoms.

In order to give the scientific name for a certain species, the genus name is given first, with its first letter capitalized; the species name, given second, is entirely in lower case. The entire name is underlined or italicized. For example, the scientific name of the cat is *Felis domestica* and that of the dog is *Canis familaris*. The cat and dog both belong to the class of mammals and the phylum of chordates.

An example of a complete taxonomic classification for a Manx cat is:

Kingdom—Animalia

Phylum—Chordata

Subphylum—Vertebrata

Class—Mammalia

Subclass—Eutheria

Order—Carnivora

Family—Felidae

Genus—Felis

Species—domestica

Variety—manx

As this example shows, important divisions may exist within a class or phylum, and the use of subclasses and subphyla is an aid to classification.

Note that the group "variety" allows us to refer exactly to the type of cats we are considering—not Siamese cats, not Persian cats, but Manx cats.

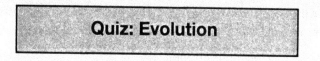

Quiz: Evolution

1. Darwin's theory of natural selection includes all of the following stipulations EXCEPT

    (A) every organism produces no more organisms than can survive.

    (B) due to competition, not all organisms survive.

    (C) the difference in survivability is due to variations between organisms.

    (D) variation is due, at least in part, to mutations.

2. The evolutionary concept that variation was due to use and disuse of parts and the concept of inheritance of acquired characteristics is associated with

    (A) Darwin.

    (B) Aristotle.

    (C) Lamarck.

    (D) Wallace.

3. The percentage of occurrence of a particular gene in a population is known as the gene

    (A) frequency.

    (B) pool.

    (C) number.

    (D) total.

4. All of the following are part of the theory of evolution originally proposed by Darwin and Wallace EXCEPT

    (A) natural selection.

    (B) inheritance and variation.

    (C) overproduction of offspring.

    (D) use and disuse of organs.

5. The random establishment of nonadaptive, bizarre types in small populations is known as

    (A) genetic drift.

    (B) migration.

    (C) selection pressure.

    (D) the Hardy-Weinberg law.

6. The most recent theories of the origin of life include all of the following elements in the primitive atmosphere EXCEPT

    (A) oxygen.

    (B) ammonia.

    (C) methane.

    (D) carbon dioxide.

7. Evolutionary convergence is the development of

    (A) dissimilar structures in organisms of dissimilar ancestry.

    (B) similar structures in organisms of dissimilar ancestry.

    (C) similar structures in organisms of similar ancestry.

    (D) dissimilar structures in organisms of similar ancestry.

8. A definition of a species is that it is a group of populations all sharing the same

    (A) habitat.

    (B) mutations.

    (C) gene pool.

    (D) acquired characteristics.

9. Choose the correct statement.

    (A) Evolution always proceeds from the simplest to the more complex.

    (B) Evolution occurs at a steady rate.

    (C) Evolution occurs at the same rate in all types of organisms.

    (D) Most new species evolve from relatively simpler forms rather than from the more advanced form of an established species.

10. The frequency of a gene in a population will NOT be largely affected by

    (A) mutation.

    (B) small population size.

    (C) selective migration.

    (D) random mating.

ANSWER KEY

1. (A)

2. (C)

3. (A)

4. (D)

5. (A)

6. (A)

7. (B)

8. (C)

9. (D)

10. (D)

CHAPTER 17

Ecology

17.1 Growth of Populations

Ecology can be defined as the study of the interactions between groups of organisms and their environment. The term autecology refers to studies of individual organisms or populations of single species and their interactions with the environment. Synecology refers to studies of various groups of organisms that associate to form a functional unit of the environment.

A population has characteristics which are a function of the whole group and not of the individual members; these are population density, birth rate, death rate, age distribution, biotic potential, rate of dispersion and growth form.

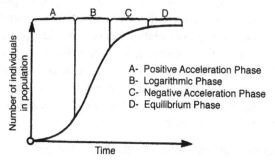

A- Positive Acceleration Phase
B- Logarithmic Phase
C- Negative Acceleration Phase
D- Equilibrium Phase

Figure 17.1 A typical S-shaped growth curve of a population in which the total number of individuals is plotted against time

A) **Population Density** – The number or mass of individuals per area or volume of habitable space.

B) **Maximum Birth Rate** – This is the largest number of organisms that could be produced per unit time under ideal conditions when there are no limiting factors.

C) **Minimum Mortality** – This is the number of deaths which would occur under ideal conditions, death due to old age.

D) **Biotic Potential (Reproductive Potential)** – The ability of a population to increase in numbers when the age ratio is stable and all environmental conditions are optimal.

Problem Solving Example:

Explain how density dependent controls of population occur.

An important characteristic of a population is its density, which is the number of individuals per unit area. A more useful term to ecologists is ecologic density, which is the number of individuals per habitable unit area. As the ecologic density of a given population begins to increase, there are regulatory factors that tend to oppose the growth. These regulating mechanisms operate to maintain a population at its optimal size within a given environment. This overall process of regulation is known as the density dependent effect.

Predation is an example of a density dependent regulator. As the density of a prey species rises, the hunting patterns of predators often change so as to increase predation on that particular population of prey. Consequently, the prey population decreases; the predators are then left with less of a food resource, and their density subsequently declines. The effect of this is often a series of density fluctuations until an equilibrium is reached between the predator and prey populations. Thus, in a stable predator-prey system, the two populations are actually regulated by each other.

Emigration of individuals from the parent population is another form of density dependent control. As the population density increases, a larger number of animals tend to move outward in search of new sources of food. Emigration is a distinctive behavior pattern acting to disperse part of the overcrowded population.

Competition is also a density dependent control. As the population density increases, the competition for limited resources becomes more intense. Consequently, the deleterious results of unsuccessful competition such as starvation and injury become more and more effective in limiting the population size.

Physiological as well as behavioral mechanisms have evolved that help to regulate population growth. It has been observed that an increase in population density is accompanied by a marked depression in inflammatory response and antibody formation. This form of inhibited immune response allows for an increase in susceptibility to infection and parasitism. Observation of laboratory mice has shown that as population density increases, aggressive behavior increases, reproduction rate falls, sexual maturity is impaired, and the growth rate becomes suppressed. These effects are attributable to changes in the endocrine system. It appears that the endocrine system can help regulate and limit population size through control of both reproductive and aggressive behavior. Although these regulatory mechanisms have been demonstrated with laboratory mice, it is not clear to what extent they operate in other species.

17.2 Energy Flow

The energy cycle starts with sunlight being utilized by green plants on earth. The electromagnetic energy of sunlight is transformed into potential energy stored in chemical bonds in green plants. The potentail energy stored in chemical bonds in green plants. The potential energy is released in cell respiration and is used in various ways.

Some of the food synthesized by green plants is broken down by the plants for energy, releasing carbon dioxide and water. Bacteria and fungi break down the bodies of dead plants, using the liberated energy for their own metabolism. Carbon dioxide and water are then released and recycled.

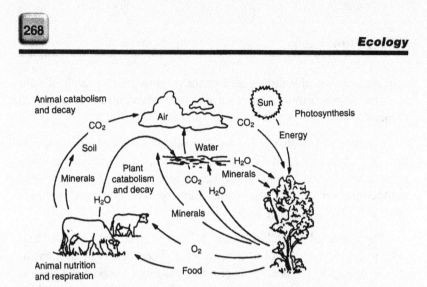

Figure 17.2 The energy cycle. This diagram shows the relationships between plants and animals and the nonliving materials of the earth. The energy of the sunlight is the only thing that is not returned to its source.

Problem Solving Example:

Q Sunlight is the ultimate energy source on earth. Energy from sunlight is not returned to its source but is transformed to other forms of energy which are closely tied together in an energy cycle. Describe the energy cycle.

A The energy cycle starts with sunlight being utilized by green plants on earth. The radiant energy of sunlight is transformed into potential energy stored in chemical bonds in green plants. The chemical bonds are synthesized by the process of photosynthesis. The potential energy is released in cell respiration and is used in various ways. Thus, fundamental to the energy cycle is the ability of energy to be transformed.

Inorganic (nonliving) matter in the ecosystem is closely tied to organic (living) matter in the energy cycle. For example, photosynthesis requires carbon dioxide from the air and water and minerals from the soil to occur besides sunlight. These are the nonliving components of photosynthesis. Chlorophyll in green plants captures the sunlight, and organic substances (i.e., glucose) are generated from inorganic ingredients via a series of enzymatic reactions in the plant cell. Chlorophyll, enzymes, and other cellular components form the living part of photosynthesis.

In the energy cycle, some of the food synthesized by green plants is broken down by the plants for energy, and consequently carbon dioxide and water are released. These again become available for green plants in capturing more energy of the sunlight. Some of the synthesized compounds are used in building the bodies of the plants and are hence stored as potential energy until the plants die. The bacteria and fungi of decay break down the bodies of the dead plants, using the liberated energy for their own metabolism. In these processes, carbon dioxide and water are released, and the minerals go back into the soil. These substances are thus recycled. Animals which feed on plants utilize a part of the energy from the food in cell respiration, with a release of carbon dioxide and water which are again recycled. Some of the minerals in the plant food are excreted by the animals and are thus available to be reused. Animals which feed on other animals utilize some energy from the food in building their own bodies. They break down some of their stored food to yield energy for daily activities such as locomotion. Food degradation is accompanied by release of carbon dioxide and water which are returned to the ecosystem. When animals die, their bodies decay and all of the materials that were used in the construction are restored to a state which can be reused by the action of the bacteria and fungi of decay.

It must be remembered that at no point in the cycle is energy destroyed. The energy from sunlight is not destroyed but is transformed into heat, chemical, or mechanical energy.

17.3 Food Web

A) Every food web begins with the autotrophic organisms (mainly green plants) being eaten by a consumer. The food web ends with decomposers, the organisms of decay, which are bacteria and fungi that degrade complex organic materials into simple substances which are reusable by the producers (green plants).

Herbivores consume green plants, and may be acted upon directly

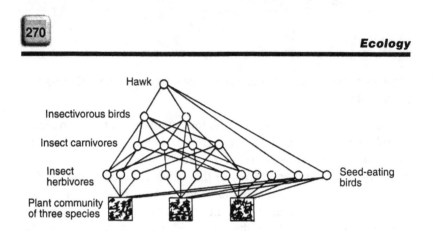

Figure 17.3 Hypothetical food web. It is assumed that in this community there are three hundred species of plants, ten species of insect herbivores, two bird herbivores, two bird insectivores, and one hawk. In a real community, there would not only be more species at each trophic level, but also many animals that feed at more than one level, or that change levels as they grow older. Some general conclusions emerge from even an oversimplified model like this. There is an initial diversity introduced by the number of plants, which is multiplied at the plant-eating level. At each subsequent level the diversity is reduced as the food chains converge.

by the decomposers or fed upon by secondary consumers, the carnivores. The successive levels in the food webs of a community are referred to as trophic levels.

B) A pyramid with the producers at the base and the primary consumers at the apex can show how energy is being supplied by the producers. It can also show a decrease of energy from the base of the apex accompanied by a decrease in numbers of organisms.

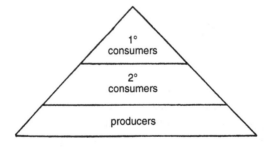

Figure 17.4 Pyramid of energy and numbers

Problem Solving Example:

Why are autotrophic organisms necessary for the continuance of life on earth? Do all autotrophs require sunlight?

Autotrophic organisms have the capacity to generate all needed energy from inorganic sources. Heterotrophic organisms can only utilize the chemical energy present in organic compounds. There are two main types of autotrophs—photosynthetic organisms and chemosynthetic organisms. Photosynthetic autotrophs obtain energy from sunlight and convert the radiant energy of sunlight to the chemical energy stored in the bonds of their organic compounds. Green plants obtain CO_2 from the atmosphere and minerals and water from the soil. Algae and photosynthetic bacteria absorb dissolved CO_2, water, and minerals through their cell membranes. Using energy from sunlight, the photosynthetic autotrophic organism converts CO_2, water, and minerals into all the constituents of the organism. Chemosynthetic organisms are much less common than photosynthetic organisms and are always bacteria. Chemosynthetic bacteria do not require sunlight and obtain energy by oxidizing certain substances. Two examples are the nitrifying bacteria which oxidize ammonia to nitrate (NO_2^-) or nitrates (NO_3^-), and the sulfate bacteria, which oxidize sulfur to sulfates. The energy released from these chemical reactions is converted to a form of chemical energy utilized by the organism.

All the organisms which carry on respiration, that is, oxidize organic compounds to carbon dioxide and water, require oxygen. Respiration is the process by which heterotrophs obtain energy. Chemosynthetic bacteria also require oxygen in order to carry out oxidations of inorganic substances. The only sources of oxygen on earth are the photosynthetic autotrophs. These organisms convert CO_2 and water to organic compounds, utilizing sunlight to provide energy and generating O_2 in the process. If there were no green plants or photosynthetic marine organisms, the oxygen present in the atmosphere would quickly be used up by animals, bacteria, and fungi.

The autotrophs are also responsible for providing the heterotrophs with organic nourishment. Sunlight is the most important source of energy on the earth, and it is only the photosynthetic autotrophs which can utilize this energy, converting it to chemical energy in organic compounds. Heterotrophs utilize the organic compounds produced by the autotrophs. Heterotrophs which obtain organic nourishment from other heterotrophs are also ultimately dependent upon autotrophs for nourishment, because the animals which are the prey have either directly or indirectly (through another animal) utilized the organic material of plants or algae. Photosynthetic autotrophs provide the earth with an energy source for living organisms and with oxygen. If photosynthetic autotrophs were not present, all life on earth would eventually cease as the food and O_2 would become depleted.

17.4 Definitions

A) **Competition** – the active demand of two or more organisms for a common vital resource.

B) **Contest Competition** – the active physical confrontation between two organisms which allows one to win the resources.

C) **Scramble Competition** – the exploitation of a common vital resource by both species.

D) **Mutualism** – a type of relationship where both species benefit from one another. An example is nitrogen-fixing bacteria that live in nodules in the roots of legumes.

E) **Commensalism** – a relationship between two species in which one species benefits while the other receives neither benefit nor harm; for example, epiphytes grow on the branches of forest trees.

F) **Parasitism** – a relationship where the host organism is harmed. Parasites can be classified as external or internal.

G) **Biomass** – the total mass of the living material present in a certain category.

H) **Succession** – a fairly orderly process of changes of communities in a region. It involves replacement of the dominant species within a given area by other species.

I) **Habitat** – the physical area where an organism lives.

J) **Niches** – include the habitat and the functional role the organism plays as the member of a community, its trophic position, and its position in the gradients of temperature, moisture, and pH.

Problem Solving Examples:

What is mutualism? Give three examples of this type of relationship.

Mutualism (or symbiosis), like parasitism and commensalism, occurs widely through most of the principal plant and animal groups and includes an astonishing diversity of physiological and behavioral adaptations. In the mutualistic type of relationship, both species benefit from each other. Some of the most advanced and ecologically important examples occur among the plants. Nitrogen-fixing bacteria of the genus Rhizobium live in special nodules in the roots of legumes. In exchange for protection and shelter, the bacteria provide the legumes with substantial amounts of nitrates which aid in their growth.

Another example of plant mutualism is lichens. They are actually "compound" organisms consisting of highly modified fungi that harbor blue-green and green algae among their hyphae (filaments). Together the two components form a compact and highly efficient unit. In general, the fungus absorbs water and nutrients and forms most of the supporting structure, while the algae provides nutrients for both components via photosynthesis.

In a different form of mutualism, many kinds of ants depend partly or wholly upon aphids and scale insects for their food supply. They milk aphids by stroking them with their forelegs and antennae. The aphids respond by excreting droplets of honeydew, which is simply partly digested plant sap. In return for this sugar-rich food, the ants protect their symbiants from parasitic wasps, predatory beetles, and other natural enemies.

In humans, certain bacteria that synthesize vitamin K live mutualistically in the human intestine which provides them with nutrients and a favorable environment.

Explain what is meant by an animal's ecological niche, and define competitive exclusion.

Ecologically, niche is defined as the functional role and position of an organism within its ecosystem. The term niche should not be confused with habitat, which is the physical area where the organism lives. The characteristics of the habitat help define the niche but do not specify it completely. Each local population of a particular species has a niche that is defined with many variables. Each has a temperature range along with other climatic factors. There are required nutrients and specific activities that also help characterize the niche.

The principle of competitive exclusion states that unless the niches of two species differ, they cannot coexist in the same habitat. Two species of organisms that occupy the same or similar ecologic niches in different geographical locations are termed ecological equivalents. The array of species present in a given type of community in different biogeographic regions may differ widely. However, similar ecosystems tend to develop wherever there are similar physical habitats. The equivalent functional niches are occupied by whatever biological groups happen to be present in the region. Thus, a savannah-type vegetation tends to develop wherever the climate permits the development of extensive grasslands, but the species of grass and the species of animals feeding on the grass may differ significantly in various parts of the world.

17.5 Types of Habitats and Biomes

There are four major habitats: marine, estuarine, fresh water, and terrestrial. A biome is a large community characterized by the kinds of plants and animals present.

Types of Biomes:

A) **The Tundra Biome** – A tundra is a band of treeless, wet, arctic grassland stretching between the Arctic Ocean and polar ice caps and the forests to the south. The main characteristics of the tundra are low temperatures and a short growing season.

B) **The Forest Biomes**

1) The northern coniferous forest stretches across North America and Eurasia just south of the tundra. The forest is characterized by spruce, fir, and pine trees and by such animals as the wolf, the lynx, and the snowshoe hare.

2) The moist coniferous forest biome stretches along the west coast of North America from Alaska south to central California. It is characterized by great humidity, high temperatures, high rainfall and small seasonal ranges.

3) The temperate deciduous forest biome was found originally in eastern North America, Europe, parts of Japan and Australia, and the southern part of South America. It is characterized by moderate temperatures with distinct summers and winters and abundant, evenly distributed rainfall. Most of this forest region has now been replaced by cultivated fields and cities.

4) The tropical rain forests stretch around low lying areas near the equator. Dense vegetation, annual rainfall of 200 cm or more, and a tremendous variety of animals characterize this area.

C) **The Grassland Biome** – The grassland biome usually occupies the interiors of continents such as the prairies of the western United States, Argentina, Australia, southern Russia, and Siberia. Grasslands are characterized by rainfalls of about 25 to 75 cm per year, and they provide natural pastures for grazing animals.

D) **The Chaparral Biome** – The chaparral biome is found in California, Mexico, the Mediterranean, and Australia's south coast. It is characterized by mild temperature, relatively abundant rain in winter, very dry summers and trees with hard, thick evergreen leaves.

E) **The Desert Biome** – The desert is characterized by rainfall of less than 25 cm per year and sparse vegetation that consists of greasewood, sagebrush, and cactus. Such animals as the kangaroo rat and the pocket mouse are able to live there.

F) **Marine Life Biome** – Although the saltiness of the open ocean is relatively uniform, the concentration of phosphates, nitrates and other nutrients varies widely in different parts of the sea and at different times of the year. All animals and plants are represented except amphibians, centipedes, millipedes, and insects. Life may have originated in the intertidal zone of the marine biome, which is the zone between the high and low tide.

The marine biome is made up of four zones:

1) The intertidal zone supports a variety of organisms because of the high and low tide.

2) The littoral zone is beyond the intertidal zone. It includes many species of aquatic organisms especially producers.

3) The open sea zone - The upper layer of this zone supports a tremendous amount of producers and therefore many consumers. The lower layer though supports only a few scavengers and their predators.

4) The ocean floor contains bacteria of decay and worms.

G) **Freshwater Zones** – Freshwater zones are divided into standing water—lakes, ponds and swamps—and running water—rivers, creeks, and springs. Freshwater zones are characterized by an assortment of animals and plants. Aquatic life is most prolific in the littoral zones of lakes. Freshwater zones change much more rapidly than other biomes.

Problem Solving Example:

 Why isn't a biome map of the earth a true representation of the vegetation formations found on land?

 The naturalist travelers of the eighteenth and nineteenth centuries saw strikingly different kinds of vegetations parceled out into formations. They drew maps showing the extent of each formation. However, drawing a map involves drawing boundaries. Where there are real, distinct boundaries on the ground this is an easy task.

Such boundaries include those between sea and land, along deserts, or beside mountain ranges. But often there are no real boundaries on the ground. In fact, most of the formation boundaries of the earth are not distinct. Real vegetation types usually grade one into another so that it is impossible to tell where one formation ends and another begins.

The apparent distinctness of the vegetation formations, when viewed from afar, is largely an optical illusion as the eye picks out bands where individual species are concentrated. A vegetation mapmaker could plot the position of a tree line well enough if he had a few reports from places with distinct treelines which he could plot and link up. However, the gradual transitions, such as between the temperate woods and tropical rain forests would be much more difficult to plot. So mapmakers draw boundaries using what seems to be the middle of the transitions. When the map is finished, the area is parceled out by blocks of formations of plants. A map like this shows the vegetation of the earth to be more neatly set into compartments than it really is.

Quiz: Ecology

1. In a food pyramid, most of the energy is concentrated at the level of

   (A) the tertiary consumer, since it is a big meat-eater.

   (B) the producer, since it supports all the consumers at the higher levels.

   (C) the primary consumer, since less energy is required to digest vegetarian meals.

   (D) the secondary consumer, since it only has to chase small herbivores and thus does not waste energy hunting.

2. Which of the following would you expect to see in a community?

   (A) About 100 times as many producers as secondary consumers.

   (B) About 100 times as many secondary consumers as producers.

   (C) An equal number of primary and secondary consumers.

   (D) An equal number of producers and secondary consumers.

3. If DDT were sprayed to kill flying insects, which organism other than the insects themselves would you expect to be most likely to suffer population losses?

   (A) fish

   (B) rabbits

   (C) birds

   (D) algae

4. As one moves up an ecological pyramid, generally

    (A) the biomass increases.

    (B) photosynthesis increases.

    (C) energy levels decrease.

    (D) the number of organisms increases.

5. The ecological unit composed of organisms and their physical environment is the

    (A) niche.

    (B) population.

    (C) ecosystem.

    (D) community.

6. The orderly change from one ecological community to another in an area is called

    (A) convergence.

    (B) climax.

    (C) dispersal.

    (D) succession.

7. Organisms which break down the compounds of dead organisms are called

    (A) phagotrophs.

    (B) parasites.

    (C) saprotrophs.

    (D) producers.

8. Which must be present in an ecosystem if the ecosystem is to be maintained?

    (A) Producers and carnivores

    (B) Producers and decomposers

    (C) Carnivores and decomposers

    (D) Herbivores and carnivores

9. The climax organism growing above the tree line on a mountain would be the same as the climax organism found in the

    (A) taiga.

    (B) tundra.

    (C) tropical forest.

    (D) desert.

10. Choose the correct statement.

    (A) Freshwater zones change more rapidly than other biomes.

    (B) The intertidal zone is between salt and fresh water.

    (C) Most of the producers in the marine biome are in the littoral zone.

    (D) The ocean floor does not contain detritus bacteria.

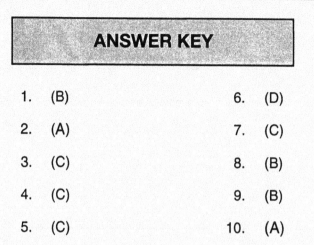

ANSWER KEY

1.	(B)	6.	(D)
2.	(A)	7.	(C)
3.	(C)	8.	(B)
4.	(C)	9.	(B)
5.	(C)	10.	(A)

NOTES

NOTES

NOTES

NOTES

NOTES

NOTES

NOTES

NOTES

NOTES

NOTES

NOTES

REA's Study Guides

Review Books, Refreshers, and Comprehensive References

Problem Solvers®
Presenting an answer to the pressing need for easy-to-understand and up-to-date study guides detailing the wide world of mathematics and science.

High School Tutors®
In-depth guides that cover the length and breadth of the science and math subjects taught in high schools nationwide.

Essentials®
An insightful series of more useful, more practical, and more informative references comprehensively covering more than 150 subjects.

Super Reviews®
Don't miss a thing! Review it all thoroughly with this series of complete subject references at an affordable price.

Interactive Flashcard Books®
Flip through these essential, interactive study aids that go far beyond ordinary flashcards.

Reference
Explore dozens of clearly written, practical guides covering a wide scope of subjects from business to engineering to languages and many more.

*For our complete title list,
visit www.rea.com*

Research & Education Association